T0275651

ENERGY, RESOURCES
AND WELFARE

Other Books by the Author:

Democracy and Sense – alternatives to financial crises and political small-talk, 2015/16.

Solar Energy Storage (ed.), 2015.

Energy Intermittency, 2014.

Artwork, 2014.

Physics Revealed Book 1: Physics in Society, 2014.

A History of Energy. Northern Europe from the Stone Age to the Present Day, 2011/2012.

Hydrogen and Fuel Cells, 2nd ed., 2011/2012 (1st ed., 2005).

Life-Cycle Analysis of Energy Systems: From Methodology to Applications, 2011.

Renewable Energy Reference Book Set (ed., 4 volumes of reprints), 2010.

Renewable Energy – Physics, Engineering, Environmental Impacts, Economics and Planning, 4th ed., 2010 (Previous editions 1979, 2000, and 2004).

Renewable Energy Focus Handbook (with Breeze, Storvick, Yang, Rosa, Gupta, Doble, Maegaard, Pistoia and Kalogirou), 2009.

Renewable Energy Conversion, Transmission and Storage, 2007.

Life-Cycle Analysis of Energy Systems (with Kuemmel and Nielsen), 1997.

Blegdamsvej 17, 1989.

Superstrenge, 1987 (reprinted 2001 and 2014).

Fred og frihed, 1985.

Fundamentals of Energy Storage (with Jensen), 1984.

Energi for fremtiden (with Hvelplund, Illum, Jensen, Meyer and Nørgård), 1983.

Energikriser og Udviklingsperspektiver (with Danielsen), 1983.

Skitse til alternativ energiplan for Danmark (with Blegaa, Hvelplund, Jensen, Josephsen, Linderoth, Meyer and Balling), 1976.

More information on the author's work is available at energy.ruc.dk, www.secantus.dk, and www.amazon.com/author/Sorensen

ENERGY, RESOURCES AND WELFARE
Exploration of Social Frameworks for Sustainable Development

BENT SØRENSEN
Roskilde University
Department of Environmental, Social and Spatial Change,
Roskilde, Denmark

ELSEVIER

AMSTERDAM • BOSTON • HEIDELBERG • LONDON
NEW YORK • OXFORD • PARIS • SAN DIEGO
SAN FRANCISCO • SINGAPORE • SYDNEY • TOKYO
Academic Press is an imprint of Elsevier

Academic Press is an imprint of Elsevier
125 London Wall, London EC2Y 5AS, UK
525 B Street, Suite 1800, San Diego, CA 92101-4495, USA
50 Hampshire Street, 5th Floor, Cambridge, MA 02139, USA
The Boulevard, Langford Lane, Kidlington, Oxford OX5 1GB, UK

Notices
Knowledge and best practice in this field are constantly changing. As new research and experience broaden our understanding, changes in research methods, professional practices, or
medical treatment may become necessary.

Practitioners and researchers must always rely on their own experience and knowledge
in evaluating and using any information, methods, compounds, or experiments described
herein. In using such information or methods they should be mindful of their own safety
and the safety of others, including parties for whom they have a professional responsibility.

To the fullest extent of the law, neither the Publisher nor the authors, contributors, or editors, assume any liability for any injury and/or damage to persons or property as a matter
of products liability, negligence or otherwise, or from any use or operation of any methods,
products, instructions, or ideas contained in the material herein.

British Library Cataloguing-in-Publication Data
A catalogue record for this book is available from the British Library

Library of Congress Cataloging-in-Publication Data
A catalog record for this book is available from the Library of Congress

ISBN: 978-0-12-803218-3

For information on all Academic Press publications
visit our website at http://elsevier.com/

Working together
to grow libraries in
developing countries

www.elsevier.com • www.bookaid.org

Publisher: Joe Hayton
Acquisition Editor: Raquel Zanol
Editorial Project Manager: Mariana Kühl Leme
Editorial Project Manager Intern: Ana Claudia A. Garcia
Production Project Manager: Kiruthika Govindaraju
Marketing Manager: Louise Springthorpe
Cover Designer: Maria Inês Cruz

Contents

About the Author

Bent Sørensen's research is cross-disciplinary and has resulted in nearly 1000 scientific articles and around 30 books, including foundation work in economic theory (the scenario method, life cycle analysis) and in energy research (renewable energy resources, technology, and applications). He has worked at universities in Japan, France, Denmark, Australia, and the United States (Berkeley and Yale), has been a consultant to governments and international organisations, a lead author in the Intergovernmental Panel on Climate Change climate assessment and recipient of several international prizes and honours.

Preface

This book is about resources and welfare, but there is a reason that the particular resource energy is singled out from the remainder. During the 1970s, energy was the asset that rose from oblivion to become a topic of vivid social debate, including discussions on the various merits of fossil, nuclear, and renewable energy supplies, the efficiency of each step in our energy conversion, and the environmental impacts of energy use, from air pollution to global warming. Also today, the transparency of energy issues, relative to many other issues in social and political debate, makes them ideal as points of departure for exploring issues of policy (that have all too long been left to professional politicians) and economic development (often optimistically restricted to 'growth', a topic that has been left for too long to professional economists).

With increasing strength, voices are heard that question the ability of the present economic and democratic paradigms and the organization built upon them to cope with the challenges of the current epoch in human history. Ordinary citizens feel excluded from exercising their democratic rights to influence politics, due to the organization of governance by way of political parties that no longer seem to span the positions held by the people that are making up the underlying societies, delivering members of parliaments that are elected with the help of advertising campaigns by a profession of spin doctors and special-interest lobbyists, and who constitute a profession of life-long politicians often caring more about re-election, than about the welfare of the societies they should serve.

This exploratory exercise will try to clarify the definitions of the various concepts entering into the frameworks constructed for discussing the issues and propose alternative ways of dealing with them. The author does not believe that there is a single solution to the problems that is superior to all other solutions, and as a consequence, several ways of dealing with the problems identified will be presented. However, what is clear is that continuing along the road laid out by the presently-implemented paradigms of social organization will not lead to a sustainable world. Other ways are needed and the longer we wait, the more difficult will it be to change tracks.

Bent Sørensen, Gilleleje, November 2015

1

Introduction

Energy has played a very special role in many social debates that have taken place over the last 50 years. Examples include the debates on the use of nuclear power, major sources of air pollution, greenhouse gas emissions and climate change and the possibility of a transition away from finite fossil resources to a sustainable system of energy supply. Quite often the topical debates on energy issues have stumbled over quite general questions on the role of private and public institutions and the policies behind the ongoing development in our societies. This catalytic effect from the specific to the general may appear surprising, but clues to the reasons for its appearance may include on one hand that energy is offering a finite set of issues much more accessible for public debate (and thus easier to start getting involved in before generalizing to a broader scope), and on the other hand the recurrent experience of sustainable solutions being blocked by the existing rules of economic and political thinking, making people reflect on the possibility of arranging society and its money flows in ways more amenable to democratic control and change.

These reflections have a particular relevance to people like the author of this book, who has spent a lifetime in research related to energy issues covering nearly the entire catalog of energy options. The question of economic assessment has come up again and again, for example in connection with comparing renewable and fossil energy systems. For fossil energy plants, the capital cost is often small and the main expense is the fuels that have to be purchased throughout the lifetime of the plant. For renewable energy plants, the up-front capital costs dominate, and only by using a present-value comparison is a fair evaluation of the relative merits of the two types of energy system possible. Most decision-makers have learned this but there is a snag: the present-value calculation depends on the interest rate employed and it is of course unknown for the future part of the system life. For this reason, an overestimate is often employed 'to be on the safe side', implying that the economy of the renewable energy system will appear less attractive than it is.

Energy, Resources and Welfare. http://dx.doi.org/10.1016/B978-0-12-803218-3.00001-X

Moving to demand-side energy investments the situation is even worse, because many consumers do not know how to compare purchases serving the same purpose but having different service lifetimes. For this reason, many consumers will fall for the cheaper product, typically produced in a developing country, although only the sales price is cheaper, not the levelized cost over identical periods of use. Typically, the lifetime of a product made in a developing economy is three to eight times shorter than that of a similar product from an established producer in a long-time industrialized country, and thus requiring three to eight successive purchases versus just one. The real cost comparison would therefore in nearly all cases favor the product with a higher initial outlay, whereas the economic interest of the industry and its commercial sales outlets typically is the opposite one.

It is a fundamental shortcoming of our educational system that primary schools have not taught children to calculate life-cycle prices and compare purchases with different service lives but serving the same purpose. The examples above derive from cases where better knowledge of a small segment of known economic procedures could have solved the problem. However, there are also cases, where the current economic paradigm is not just being misused by amateurs but faces fundamental shortcomings. The following chapters will address some of these and discuss various ways out of the problems. First the current paradigm will be described in some detail, then historical and new proposals for remedying the situation will be presented, together with identifying obstacles for implementing some of the solutions. Finally, a scenario combining all the ingredients will be outlined for the reader's consideration, reflection and possibly criticism.

The investigation of the role of economic paradigms in society will necessarily lead to excursions in several directions other than economic theory, including discussion of physical and intellectual resources and lifestyles and social organization at both local, regional, and global scale. Concrete examples or case studies of past dealings with the issues will be presented, including a fairly detailed discussion of the energy use in specific sectors. Vague concept such as "growth" will have to be disassembled into components characterized by having positive, neutral, or negative impacts, and the entire range of human activities will be reassessed in light of general considerations of sustainability and of the distribution of benefits and burdens, leading to discussions of disparity in welfare, political power structures and tolerance toward variations in human preferences and choices. The bottom line, as first emphasized by ecologists, is that everything seems connected and that partial solutions may not lead to optimization of the fate of the entire human community embedded into the constraint of a finite, small planet.

The ambition to "solve" all the problems identified during this odyssey is of course impossible, although trying is not necessarily the worst approach. A more attainable goal may be to create a framework for a qualified debate of the issues raised, which can inspire others to contribute to the many gaps in the arguments of this modest attempt to flag the problems. This is certainly felt to be needed, considering that professional politicians in most parts of the world do not spend any time questioning the neoliberal paradigm underlying the running of nearly all societies of the present world (whether denoted conservative or socialist/labor), and despite voices from an increasing mass of knowledgeable people expressing their doubts. My aim has been to get to the roots of these issues, which has led me to question whether we actually have democracies beyond the mere use of the name and whether the basic rules presently embedded into constitutions and declarations of human rights are sufficient and if they are reflected in the actual execution of governance.

My suggestion is that human rights gain weight when they are connected with human duties and, in light of how governance is conducted in different parts of the world, we probably need to restart the discussion of democracy from the bottom. This includes a new consideration of the pros and cons of direct and representative democracy, where the representative democracy seen today has developed in a covert activity conducted by a small group of professional politicians, backed by a small elite group and elected by advertising campaigns carefully choosing and avoiding issues to be exposed to the voters. This often subsequently leads to ruling by massive misinformation and blocking of relevant information, thus increasingly alienating the population and blurring the distinction between democracies and other political set-ups of oligarchic or dictatorial form.

The question is raised of whether we should continue the motion toward one world, forcing trade, commerce, and other interaction to be globalized despite the different speeds of development evident in different parts of the same world. Could there not be benefits of having a different organization in different parts of the world, just as each region obviously has different views on the basic objectives and different claims on what constitutes the human predicament. If so, it would need to be sorted out, which playing rules need to be globally imposed in order to peacefully share the planet and which can be left to individual societies to shape according to their preferences and state of development.

A general audience book exploring the political implications of the analysis made in this work is available (Sørensen, 2016).

Reference

Sørensen, B., 2016. Democracy and sense – alternatives to financial crises and political small-talk. Secantus, Gilleleje, and Amazon CS.

Resources and Economic Welfare

2.1 OVERVIEW OF NATURAL RESOURCES

2.1.1 Land and Property

Land is an asset of basic importance to mankind. Hunters and gatherers depended on the presence of ranges and forests for their prey and picks of fruit and other edible plants. In a general sense, the waters used for fishing constitute a land resource, because geologically it is just land with a modest cover of water that has changed over time. During the ice ages, the global average sea levels changed by over 100 m due to the binding of water masses in ice.

Later, agriculture, fish farming, and livestock imposed the need for human societies to appropriate and control land for such purposes. This Neolithic transition presumably also formed the notion of property because having toiled at clearing and sowing a piece of land, a person felt it should be his or her possession, at least until the time for harvesting and better forever. This gave rise to the idea of inheritance, passing the property to the next generation, justified at first by the effort imbedded in the conversion of the land to agricultural standards, but later by attaching a utility-based value to the land.

Soon farming had a profound impact on the stratification of society into those possessing land and those who did not, a division made durable by the concept of inheritance. The notion of property also created the need for defending one's property against anyone suspected of wanting to steal or take over the land, so the invention of weapons and war ensued. One suggestion is that the transition from mythology (existential thinking) to religion (formation of institutions with set prescriptions for such thinking) was also happening as a result of the need felt to make some people, such as those who did not own land, accept slavery or slavery-like conditions of low welfare, provided that they were compensated by the newly-invented idea of a better afterlife (Sørensen, 2012a).

Energy, Resources and Welfare. http://dx.doi.org/10.1016/B978-0-12-803218-3.00002-1

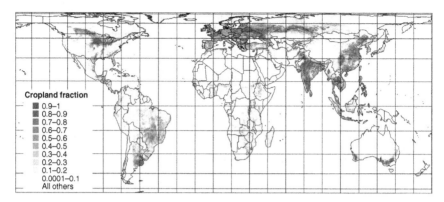

FIGURE 2.1   Cropland fractions (data downloaded from University of Columbia, 1997–2015, based on USGS classifications described in Loveland et al., 2000, and in USGS, 2015a). In the northern temperate regions, these fractions are currently declining. The plane projection used for the datasets here is the equal-area Behrman Earth projection. This means that a unit area everywhere on the map represents the same physical surface area of the Earth, although skewed.

Today, land and property rights have been further elaborated and monetary values have been assigned to each asset. This first of all signals the fundamental importance of these assets to the current structure of societies, divided as they are into nations formed upon territorial ambitions. With a negotiable price tag attached it is now possible to buy and sell land as well as other property (from buildings to equipment and claimed intellectual achievements). The price of land has been linked to its allowable uses, which in many parts of the world is governed, for example, by zoning legislation, causing average prices per unit area to lie in different cost brackets depending on whether the land areas are classified as urban, marginal/recreational, or reserved for agricultural or forestry usage. Figs 2.1 and 2.2 show current global agricultural and rangeland (for crops or livestock), Fig. 2.3 shows the forestland, of which most is exploited for timber or other wood applications and Fig. 2.4 shows the urban land, including land occupied by infrastructure, such as roads and buildings. The remaining global land area is classified as marginal or recreational. Figs 2.5 and 2.6 show the offshore waters used for capture of some major fish resources, and Figs 2.7 and 2.8 correspondingly the areas, mainly inland, used for some important aquaculture plant or fish/shellfish production. Areas used for extraction of mineral resources will be discussed in Section 2.1.2.

Agriculture has a history stretching back more than 10,000 years, and the areas used in this way have changed over time in ways considerably

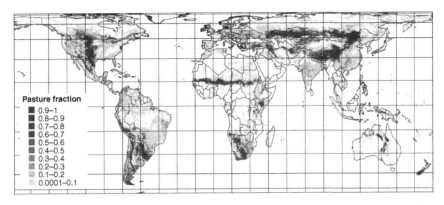

FIGURE 2.2   Pasture fractions (source: as Fig. 2.1).

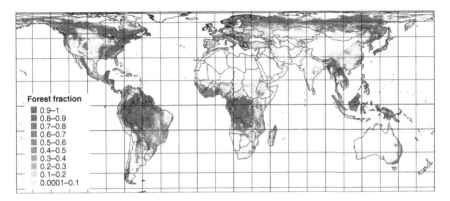

FIGURE 2.3   Forest fractions (source: as Fig. 2.1). In several southern regions, these fractions are currently declining.

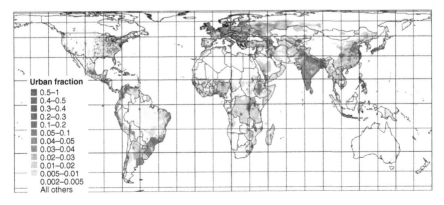

FIGURE 2.4   Urban fractions (source: as Fig. 2.1). These fractions are generally increasing.

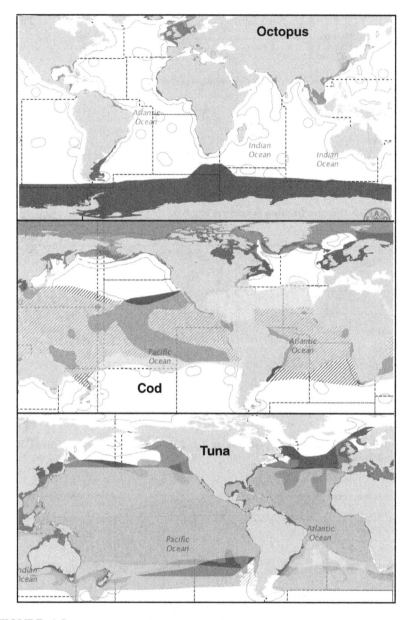

FIGURE 2.5 Geographical distributions of catch areas for selected fish species (FAO, 2015a). The hatched areas for cod are considered uncertain. A discussion of the models used to map catch and capture areas may be found in Megrey and Moskness (2009).

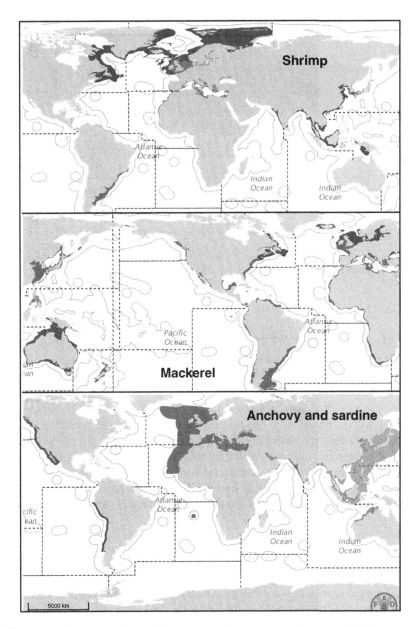

FIGURE 2.6 Geographical distributions of catch areas for selected fish species (FAO, 2015a).

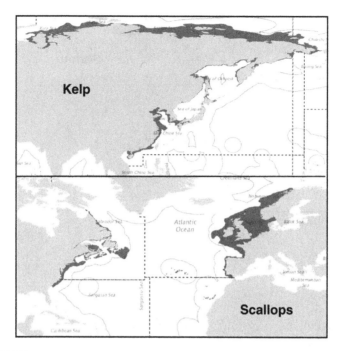

FIGURE 2.7   Geographical distribution of the use of aquaculture and offshore catch for production of kelp and scallops (FAO, 2015a).

more complex than just reflecting population variations.* The efficiency (crop yield per unit of area) has not increased systematically with time, but has been affected by both climatic and cultural changes. In the northern temperate zones, climate is favorable for agriculture, with enough precipitation but few torrential rains, and with adequate solar radiation during the growing season but few incidents of scorching by high levels of radiation. Nutrients are also available, due to the stable humus layer

*A model called HYDE referred to in many publications (Goldewijk et al., 2011) assumes a constant area of cropland per capita over time, which is clearly false. Attempts have been made to improve it by assuming increased farming yields per capita caused by improvements in the efficiency of farming, prompted by better plows and tilling equipment, as well as by better nutrient recycling, first provided by livestock rotation practices and later by administration of chemical fertilizers (Kaplan et al., 2010). However, even this model does not agree with the data available for selected areas (e.g., Sørensen, 2012a), due to the influence of cultural and climatic differences, as well as external factors such as warfare and periods with use of slaves for agricultural work, usually connected with declining yields and delayed introduction of more efficient practices. The population model used by HYDE (Goldewijk et al., 2010) is also dubious, as it is based on estimations of population distributions and changes made on the basis of speculation, at least for regions outside Europe (McEvedy and Jones, 1979).

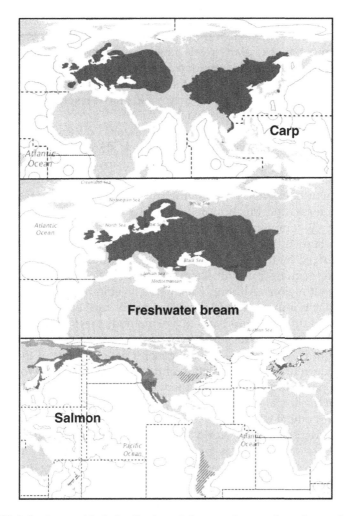

FIGURE 2.8 Geographical distribution of the use of aquaculture for production of selected species (on country basis, except for salmon; FAO, 2015a).

formed over long periods of time, with at least a partial recycling of nutrients by simultaneous soil-conditioned cultivation and spreading manure from animal husbandry. Yet the farming yields in these regions have gone up and down under the influence of land ownership and economic practices, and over long periods of time this has left agricultural work to an underpaid stratum of society, with no attempts to improve the practices by advances in the tools used. Only recently, mechanization and scientific optimization of methods have caused an unprecedented rise in productivity, so that some agricultural areas are currently being converted into fallow land in order to keep the overproduction relative to demand so low that market prices are not seriously affected. The crop yields in north European societies reached around eightfold (i.e., eight times the continuity requirement of next year's seeds for sowing) during the late Stone Age and Iron Age, but subsequently decreased to a mere twofold during the Medieval and Renaissance periods (Sørensen, 2012a). After major agricultural reforms during the 19th century, yields have resumed growth and are now around 100-fold in Denmark.

The situation in regions closer to the equator is different. Historically, domestication of both cereals and livestock started in Mesopotamia, which was then a fertile area. Today, this and many other early sites of agriculture have been transformed into dry land or desert, likely as a result of the human interference. Removing trees makes the evaporation from the soil increase, and much more than is the case in temperate climates. This process is compounded by planting annual crops that during parts of the year occupy only a fraction of the land surface. The current situation is characterized by the use of inefficient agricultural technology in many low-latitude regions, but with increasing application of irrigation in the more affluent regions. Although below the level in high-latitude regions, the efficiency is improving and there is room for further improvement using advanced drip-irrigation and tillage methods together with adequate nutrient management. This will be necessary due to expected population increases but it is clearly tied to the general development of the regions involved. Current world production of food may be adequate for feeding the actual population, but due to the variations in economic ability and the rules of the liberal market economy, there are regions where people are starving and others where surplus food is discarded as waste.

In some regions of the world the ratio of meat and vegetable components in the human diet is greater than considered healthy. However, meat and fish components play an essential role in providing the proteins and other constituents necessary for human health. A vegetarian diet can supply these necessities only by applying certain fermentation techniques to plant material growing only in few parts of the world. Particularly, fish and shellfish are often seen as an important option for feeding a growing world population, but this would only work if the catch areas can be effectively protected against overfishing that may reduce stocks of certain species. Another problem is that seas and waterways have been used to dump waste of all kinds (including radioactive substances) for quite some time, and pollution with unhealthy trace elements is even felt in the largest oceans.

During the latest decades, aquaculture has grown in importance and currently supplies vegetable matter (e.g., kelp, seaweed), shellfish (clams, shrimps, etc.) and fish (salmon, bream, etc.) in rising quantity. Poor management of fish farms has exhibited a number of risks similar to those of industrial pig and cattle farms, with epidemic diseases periodically spreading over human societies as a consequence. Some of these issues may be handled by reducing the size of individual (fish or meat) farms, and by replacing the administration of pesticides and preventive antibiotic treatment by health and environmentally safe methods, such as certified ecological practices (also called "organic" agriculture and aquaculture). The known 10–20% reduction in yields obtained by sustainable ecological farming would seem acceptable at least in the present period of surplus production, by avoiding abandonment of areas previously used for farming (Ponisio et al., 2015; Sørensen, 2012a).

Property is usually defined as assets that are accorded ownership and considered heritable. This caused severe criticism of the property concept during the period following the French Revolution (Proudhon, 1840), but had actually been discussed earlier in England, during the mid-17th century (Woodhouse, 1951). In principle the nature of the property asset can be anything, from land and buildings to intellectual achievements, such as patents or copyrighted art and literature. Although available statistical surveys, such as national accounts are keen to include private property, the treatment of public property is often scanty. This is because most statistics focus on economic valuation, and while private property is routinely valuated in order to form the basis for taxation, government property, such as the historical buildings, waterways, or the underground with whatever resources it may contain, and publicly owned protected areas and museums with art or historical/archeological artefacts are rarely subjected to a precise monetary evaluation. This makes it difficult to assess the loss to society when governments decide to privatize public assets to gain short-term liquidity, and more generally, it makes it difficult to quantify the total wealth of a society.

Wealth is a concept related to property. As a general term, wealth has appeared in historical literature for a very long time, also before Adam Smith's treatise (Smith 1776). When econometrics entered the scene, wealth received a monetary definition as the present value of future consumption (Fischer, 1906). "Present value" means that future economic payments are replaced by a current value by reverse-discounting the actual payment for future consumption, that is, replacing it by the sum of money that today would have to be set aside (deposited or invested) to generate the necessary future sums after inclusion of accumulated interest. Interest rates should not be high like those used for individual home or consumption loans (usually high due to the finite lifetime of humans – interest rates measure the advantage of having a sum of money now rather than N years into the future) and not the more modest type, sometimes called "social interest rates," of government bonds (on average around 3% per year over the 20th century), but they are supposed to be the intergenerational interest rate ensuring that future generations are not given poorer conditions than the current generation of people. Such intergenerational interest rates could have positive components from the stock of useful things we leave to the following generations, but negative components from the environmental degradation and resource exhaustion we commit. Unable to determine these components in detail, one should probably just set the intergenerational interest rate at zero. It simply cannot be said if future societies would be happy with the assets and problems we pass to them because history has taught us that especially problems not thought of when the act that caused them was committed have a habit of turning up and becoming problems at some later moment in time. Some current economists pretend to "play it safe" by assuming a positive intergenerational interest rate (1.5% per year in a study by the World Bank, 2011, also Hamilton and Liu, 2014), while others argue for zero interest on ethical grounds (e.g., Quiggin, 2012).

In any case, the stock concept of wealth has for a while been shown little interest by the economists and governments making national accounts, relative to the flow measurements of annual income and expenditure, and only recently have a number of economists called for the procurement of better data that would allow renewed focus on wealth, including its distribution among nations (notably the precursors to the World Bank study references and studies of household wealth distribution and inequality by the United Nations University, e.g., Ohlsson et al., 2006; Davies et al., 2008, 2009). National accounts have allowed detailed investigations of (private) income distributions (OECD, 2012), but wealth distributions are incomplete and highly uncertain (Piketty and Saez, 2014; Piketty, 2014). However, attempts to remedy the situations have been made by adding proxies for some of the wealth categories, for which national data are not directly available, notably by Hamilton and Clemens (1999), World Bank (2006, 2011), Liu (2011), and Hamilton and Liu (2014).

To explain these ideas the relationship between wealth and the concept of welfare (appearing in the title of this book) should first be clarified. Dasgupta (2001) defines welfare as the present value of utility along a path of development, utility being the ability to satisfy human needs or desires, a concept that does not lend itself easily to measurement. Some economists distinguish between personal welfare and social welfare, recognizing that a society may have needs (e.g., for coherence) going beyond those of the individual's needs. The focus is on evaluating (consumer) goods, for which utility would be their constant or declining usefulness over the lifetime of the product. Referring to both needs and desires makes the concepts of utility and welfare very woolly, depending on pressures from peer groups or advertising agents (advocating e.g., sports cars, designer dresses) and many economists prefer to replace them by wealth, arguing that wealth is a fair proxy for welfare. Rather than looking at absolute values, Hamilton and Clemens (1999) use the change in social welfare, which they call "genuine saving" after correcting the value of assets produced but not consumed for depreciation and for stock building, such as increased knowledge or with negative sign depletion of environmental assets. It is not considered that welfare could contain components that cannot be monetized. This is reminiscent of the situation 50 years ago, where economists called social and environmental impacts "externalities" and omitted them from their analysis, albeit sometimes expressing slight regret. Today, we routinely try in some way to incorporate "externality" questions in policymaking, no matter if they can be monetized (which has become possible, although inaccurate, e.g., for air pollution and global warming impacts) or not (e.g., biodiversity, leisure time, cultural activities, human relations).

The World Bank studies associate wealth with three types of capital:

- Produced capital (machinery, structures, equipment, etc.)
- Natural capital (agricultural land, protected areas, forests, mineral deposits, energy resources)
- Intangible capital (human, social and institutional capital plus net financial assets), calculated as total wealth minus produced and natural capital.

Wealth is, as mentioned, taken as the present value of future consumption, but in practice it is restricted to categories appearing in national accounts and tuned by assumptions on utility life and interest rate, the two parameters without which, the present value cannot be calculated. This procedure is of course prompted by the availability of annual accounts and short-term forecasts and the unavailability of proper wealth data. The intangible capital found by these studies turns out to be very uncertain (and sometimes negative). Liu (2011) identifies human capital as the most important part of intangible capital and uses years of school attendance as a proxy for it, leaving what seems a small remainder, however which may just be taken as a reminder that many things are omitted from the analysis, such as items difficult to monetize, and that crude lifetimes and discount rates are used. In the World Bank study, a 4% interest rate is used for land and underground mineral resources, including fossil fuels, with a depreciation time of 25 years or the time taken to use remaining proven reserves. For produced goods or structures (Fig. 2.9), the World Bank uses a 5% interest rate and a 20-year lifetime.[†] This covers different items, where consumer goods and industrial machinery currently have

---

[†]Produced goods are included in the discussion here, although the section is about "natural resources", because the produced goods may be regarded as processed natural resources. Of the 4 or 5% depreciation rate used by the World Bank, 1.5% is interest, the rest is said to represent consumption growth. This may make sense for depletable mineral resources, but why land value depreciation should depend in this way on consumption growth, is difficult to appreciate.

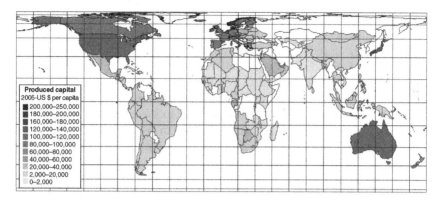

FIGURE 2.9 **Country distribution of 2005 wealth derived from the present value of produced stocks (monetized as accumulated investments whenever data allows) available for use during the depreciation period, of machinery, equipment, buildings, and structures, such as roads, bridges, and urban areas.** The depreciation period is an assumed average service life of 20 years with a 5% per year interest rate (World Bank, 2011). Depreciation to zero excludes the possibility of recycling parts of the produced stock. Countries shown in white are not included in the survey.

lifetimes of 15–20 years (or less, if they are produced in certain developing countries), while buildings have typical lifetimes of 80–200 years, and structural items, such as roads and bridges have lifetimes of 50–100 years, all depending on the level of maintenance and repair undertaken (and recorded as annual flow expenses despite their influence on product lifetime and thus wealth). The interest rate used by the World Bank seems to be a personal one reflecting the options of a private consumer for borrowing.

The natural capital included in the World Bank model of wealth include fossil energy and other mineral resources with a depreciation linked to depletion, but also cropland, forests, and protected areas (valued as if they were in the other categories). The valuation is a kind of rent,[‡] given that it depreciates over a 25-year period, and the assumed interest rate includes a part reflecting consumption growth. However, the problem is more serious because a piece of cultivated land or a forest can continue to be productive for any amount of time, provided that sound farming and forest replanting techniques are used. This means that one must use a long-term intergenerational interest rate, which, as argued, should probably be taken as zero. The argument for this is furthered by considering that the stock of building and structure assets we may pass on to future generations will last no more than at most a few hundred years, while the negative impacts of for example, radioactive waste or greenhouse gas emissions will prevail for much longer. Taking the long-term interest rate as zero implies that the value or wealth constituted by permanently used

‡It will become clear that this trick saves the World Bank study from dealing with infinities.

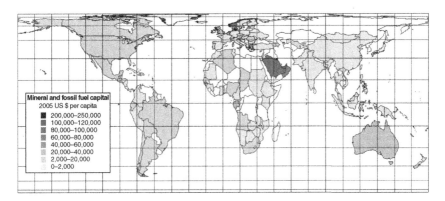

FIGURE 2.10 Country distribution of 2005 wealth derived from the present value of future use (assuming an exponential growth rate of 2.5% per year and monetized at current prices) of proven reserves of fossil fuels and some important subsoil minerals, depreciated over an assumed average life (to depletion) of 25 years with a 4% interest rate, including the 2.5% annual growth (World Bank, 2011). Possible recycling of certain minerals is not considered. Countries shown in white are not included in the survey.

land, forest, or water property is infinite. The World Bank tables showing a finite "natural capital" based only on 25-year rental of the property are simply meaningless, and the durable natural resources fundamentally cannot be treated as wealth on the same footing as consumer products. This constitutes a strong support for Proudhon's suggested abandonment of property rights, at least associated with land, forestry, fisheries, and use of renewable energy (where solar and wind power is often favorable precisely on the marginal land otherwise valued lowly). Instead, one could restrict handling of these assets to renting agreements with inheritance replaced by new rental. In fact this is not too different from the actual situation in many parts of the world, for example, where farmers only nominally own their land. In reality, they have borrowed the money and pay an annuity every year to a bank or real estate institution of private or public status. Replacing all long-term durable property ownership by rental, the wealth comments are again on the same footing because this is or could be the setup for the consumer goods of finite lifetime, making wealth a concept describing the volume of assets that a person (or a society) has been able to secure rental agreements for, presumably involving some assessment of the person or society's ability to pay for the rental, by work or otherwise. This definition of wealth avoids the infinite value of future options for consumption.

Fig. 2.10 shows the World Bank figures for wealth associated with nonrenewable mineral resources, including fossil fuels, the only components of "natural wealth" that makes some sense. As mentioned, the present value is calculated with a 25-year depreciation time (to depletion) and a 4% interest rate, reflecting 2.5% consumption growth, and a 1.5%

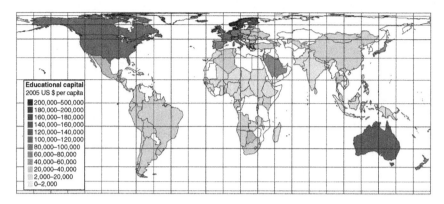

FIGURE 2.11 Country distribution of 2005 wealth associated with educational skills in the population, taken as current per capita annual expenditures on education (World Bank, 2011) times 50 (years). Educational skills would often have to be refreshed by continued education over the assumed 50 years that a person on average will offer services based on such skills, but this is considered as included in current teaching budgets and thus no depreciation of knowledge learned is applied. Countries shown in white are not included in the survey.

real interest rate. Renewable resources are not considered. They should be associated with land areas and be assigned an infinite lifetime (solar and wind energy are power-derived from conversion of energy received from the Sun, and so is hydro-derived from the global water cycle, again sustained by high-quality energy received from the Sun and returned as lower-grade energy to the universe). These renewable resources make many areas of land classified as "marginal" as valuable as cropland, forestland, or sea areas, allowing fishery or aquaculture. The accumulated wealth connected with renewable energy flows, is clearly much higher than that connected with depletable resources, such as fossil fuels.

The Fig. 2.11 shows a significant part of what the World Bank study calls "intangible capital," namely, the educational skills available in different countries, based on expenditures for the educational sector in each country, and multiplied by the number of years an average person is able use such skills productively. Generating educational skills for the next generations involve new educational efforts and expenditures, so in this case the wealth created in finite. Other components in the "intangible" category include human, social, and institutional capital, including cultural assets of science, music, literature, the arts, as well as historical and archeological artefacts and their restored synthesis. These are (hopefully[§]) never lost, and thus constitute nondiscountable wealth components. The actual World Bank assessment further includes net financial assets, which may change with time and thus affect the estimates of wealth at a particular moment in time.

[§]Bad exceptions are the Taliban and Islamic State destruction of historic treasures.

Since cultural and shared intellectual capital is durable and permanent, in contrast to education that must be renewed for each generation, evaluating them will give infinite results, like for land. Current practice is to allow ownership for a limited period (patents, copyrights) and then place such assets in the public domain. Generally, the upgrading of natural and cultural assets relative to current practice of mainstream economists discussed previously, relative to quickly consumed goods, offers an explanation of the fact that some studies find (nonquantified) indications of high welfare in societies with high public awareness of the intangible assets, but still with an emphasis on relatively recent conquests (in areas, such as free health care, free education, unemployment compensation, gender equal opportunity, and not least the full range of civil liberties encoded in the Declaration of Human Rights by the United Nations, UN, 1948).

The World Bank estimates of land value (land used for agriculture, forestry or recreation) are not shown because it replaces wealth by a 25-year rental income from these assets, while as argued the proper benefit period is nearly infinite. To this, the World Bank economists would argue that with a large, positive interest rate, the period beyond 25 years is unimportant. However, this is not a valid argument because there is no stock building to support a positive interest level. This can be seen by narrowing the arguments made earlier to land assets; no improvement of the value of the land itself can take place, only possibly increased yields due to improved agricultural practices (at a given time), which do not alter the value of the land itself (but may alter its sale price in a market). Negative long-term interest rates are possible if nutrients are removed from the land (recycling manure can never conserve 100% of the nutrients) or if use of pesticides degrade the land. Thus, a negative long-term interest rate for natural assets, such as land is likely, and the zero value is really the highest that can be defended.

Other studies of welfare have approached some of the data-poor areas by interview studies, which of course have their own set of caveats and uncertainties. Terms like "satisfaction" and "happiness" are employed, presumably to signal the subjective nature of some of the methods used (Estes, 1984; Legatum Institute, 2011). A credible approach to describing welfare in qualitative or quantified terms must address not only household or private wealth components, but also public assets, such as the total physical environment (from government-owned assets to the "commons"), the infrastructure system, health, administration, and educational systems (beyond the building and construction parts included as produced capital), as well as cultural assets involving arts, music, literature, and other media (comprising recorded deliberations on social arrangements and governance), science and other knowledge, any preserved historical or archeological artefacts, in summary embracing all that contributes to a specific civilization.

The deliberations presented in this section have shown that the seemingly innocent assumption of a zero intergenerational interest rate has far-reaching implications, including demonstration of the absurdity of private ownership and property inheritance for any piece of the natural Earth or its renewable energy flows. No one can in this perspective afford ownership of land. Land must be a common good like solar energy or wind. The economist definition of wealth and welfare in terms of a present value of future consumption takes a different shape with the recognition that some assets can sustain indefinite profit (provided that we do not abuse and destroy them) and are immune to depreciation because the equal consideration of the present and any future generation of planetary inhabitants requires the long-term interest rate to be zero. Short-term rental arrangements have not been excluded, but whether this is the best way to go or not will have to be further discussed in Chapter 3 on governance and paradigm shifts.

In any case, the remarks made here suggest a major revaluation of the capital or wealth values assigned to different property categories, with natural and cultural assets having infinitely higher value compared to the items reflecting running consumption.

2.1.2 Mineral Resources Including Energy Resources of Geological Origin

Some mineral resources are shown in Fig. 2.10, but have been evaluated in monetary terms as a kind of wealth based on their short-term potential for commercial trading. In reality, the mineral resources called "depletable" are not really "used," but transformed into other chemical forms, for which economic exploitation is less obvious. The carbon of fossil energy resources is converted into carbon dioxide, which has far fewer commercial uses (e.g., for sparkling drinking water or as a reinjection gas used to increase the recovery fraction of mining other subsurface fossil fuel deposits) than the originally extracted oil, gas, or solid coal, and which if not collected, will stay for considerable lengths of time in the atmosphere and cause increased greenhouse warming. Also mineral resources, such as metals or rare earth elements are not disappearing, but transformed and incorporated into, say, consumer goods (from window frames to computer circuits), from which they may, at the end of the product's service life, be recovered after dismantling and disassembly and reused or recycled. Currently, many produced pieces of equipment are poorly manufactured as judged from a recycling perspective, requiring much effort to disassemble and thus inviting the less efficient melting down process of incineration, followed by a partial recycling of only those components that have not escaped as air or waterway pollution but lend themselves to separation into new primary materials for a second manufacturing cycle. Of course, using

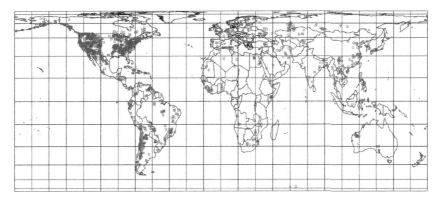

FIGURE 2.12 Survey of the global distribution of operating iron mines (USGS, 2015b; USGS warns that the database is no longer updated, and the apparent difference in the number of operations inside and outside the United States may reflect that some information outside the US is missing. However, there are also differences in the size of mining operator companies, where, for example, the United States has a tradition for many small operators. These remarks pertain to the following figures of the distribution of other minerals as well).

economic price setting as a guideline for recycling, the declining availability of virgin materials will eventually force an increase in recycling fraction. Still, even future insistence on products optimized for end-of-life disassembly will not allow recovery of minerals that have already been reduced to compounds spread over the environment in concentrations not inviting extraction.

Figs 2.12–2.16 show a selection of important mineral resources currently mined and mapped according to their location. The quantity of minerals at each site is not shown and is less easy to quantify than the resource

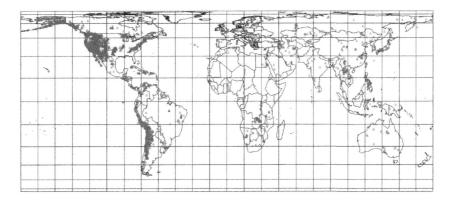

FIGURE 2.13 Survey of the global distribution of operating copper mines (USGS, 2015b).

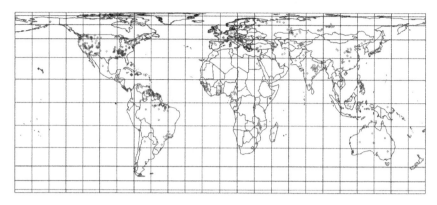

FIGURE 2.14 Survey of the global distribution of operating aluminium mines (USGS, 2015b).

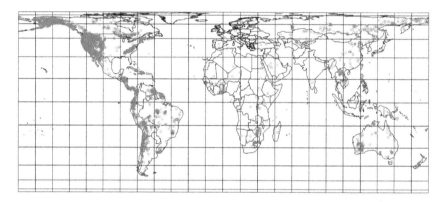

FIGURE 2.15 Survey of the global distribution of operating gold mines (USGS, 2015b).

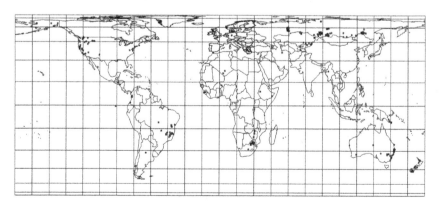

FIGURE 2.16 Survey of the global distribution of operating platinum mines (USGS, 2015b).

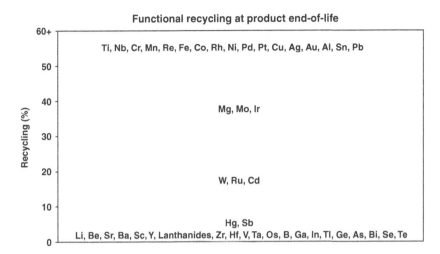

FIGURE 2.17 Functional recycling of metals at product end-of-life, i.e., recycling with retention of the physical and chemical properties of the metal in its previous use (with permission from Graedel et al., 2011).

extraction level that may be achieved in the next decades, eventually by exploitation of new deposits, either those already known or those that may be found in promising geological areas. Often, the geology of underground ores and composites do not allow a precise quantification of the overall resource or the economy of recovery. As always in resource estimation, there is a cut-off toward concentrations, for which recovery is considered highly uneconomical or impossible, at least with currently known extraction technology. One uses terms, such as proven reserves, possible reserves or ultimately recoverable resources to classify the estimates. The proven resources are often quite low, allowing only some 15 years of extraction at current level for minerals such as gold, silver, or lead and no more than 90 years for resources such as iron and aluminium (World Bank, 2011).

However, unlike fossil fuels, mineral resources like metals or building materials are not irrevocably depleted by their use in human society, but may as mentioned be reused or recycled. At present, recycling fractions range from under 1% (e.g., lithium or germanium) to over 50% (e.g., iron, copper, silver, gold, and platinum), as shown for the product end-of-life recycling levels in Fig. 2.17. Additional recycling options exist at the resource extraction and product manufacture levels. The global amount of minerals imbedded into buildings and other structures, vehicles, industrial equipment, and consumer goods on average works out to some 80 kg/cap. of aluminium, 45 kg/cap. of copper, 2200 kg/cap. of iron, 8 kg/cap. of lead, and 0.1 kg/cap. of silver, with the transportation sector being the primary user of aluminium, followed by the buildings and construction sectors, which are also the prime users of copper, iron, and lead, while the principal use of silver is in consumer goods, such as jewellery (UNEP, 2010). Some minerals are presently growing in importance, for example, lithium used in batteries that are currently expanding from consumer electronics to road vehicles, and this will certainly demand moving Li from the low recycling level of Fig. 2.17 to a much higher value (Mohr et al., 2012).

Whereas current mineral extraction is mostly from mines drilled into the subsoil of land areas, at the surface or deeper, there has long been an interest in offshore mining (Odell, 1997; Baturin, 1997), notably of the nodules with a high mineral content found on the bottoms of many ocean locations. However, the cost of extraction is in most cases still seen as too high. New focus with more short-term prospects has been directed at what is called 'urban mining', meaning extraction and recycling of mineral resources from the urban waste, which currently is increasingly being collected and separated (rather than incinerated, at the source or at a waste treatment centre) in a number of countries around the world (Zhu, 2014). A variety of recycling processes, adapted to the relevant product mix, have been developed and in many cases, and recycling rates of over 90% are available (e.g., for chromium and certain iron products; UNEP, 2011) or promised in the near future. An important example is recycling of rapidly growing amounts of scrap from electronic equipment (Tuncuk et al., 2012). For other products, such as the dominant current type of solar cells, the recycling request must be based on environmental concern because the chief material, silicon, is an abundant mineral and recycling therefore not warranted by direct economy (McDonald and Pearce, 2010).

2.1.3 Flow Resources, such as Renewable Energy

Renewable energy resources are nearly all based on conversion of radiation received from the Sun, eventually returning the energy to space as lower-quality heat. The term 'renewable' is just used to point to the extremely long period (billions of years) that the nuclear processes in the core of the Sun are expected to continue. Direct solar radiation can be converted into electricity or heat. The source flow depends on the location on the Earth, as shown in Fig. 2.18. It is largest in subtropical regions and small near the poles, in addition to increased seasonal variation. The variations in solar radiation (including day-to-night change) give rise to temperature differences, which create the pressure differences in the atmosphere that again give rise to winds. Winds further create waves at water surfaces. Less than 1% of the solar radiation is required to maintain the winds, and less that 1% of the winds are required to create the waves. However, wind energy is more concentrated than the original solar energy (typically up to five times more wind energy passes one (vertical) square

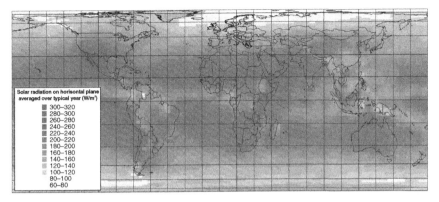

FIGURE 2.18   Annual average solar radiation on a horizontal plane (W/m²), at the surface of the Earth (ECMWF, 2015; Kållberg et al., 2007). Devices for collection of electric and/or thermal energy may be tilted toward the Equator, giving an increased yield at higher latitudes than a horizontally placed collector. Further tracking to the direction to the Sun may provide even larger output, but at an expense.

meter than solar energy falling on one (horizontal) square meter (Sørensen, 2010). Yet wind energy is available rather evenly spread over the year. Wave energy is accumulating wind energy, but is available much fewer hours over the year, implying that the cost of a wave-energy collecting device being able to withstand maximum wave impact is high compared with the annual production.

Hydropower is also based on atmospheric processes triggered by solar radiation, namely, the evaporation and condensation of water. These processes involve about a quarter of the incoming solar radiation, primarily due to the substantial energy required for evaporation. Precipitation is one of the most important flow resources, and is essential for plant growth as well as for the use of surface water resources in a sustainable mode. A small fraction of the energy in the water cycle appears as hydropower, based on the gravitational energy associated with water vapour having condensed at some elevation. Fig. 2.19 gives an impression of the available hydropower, based on water runoff measured or calculated with use of atmospheric modelling (i.e., the crude, large-scale only, circulation models used in weather forecasts and in climate change calculations) and the topological heights. Exploitation of hydropower is possible only in a fairly few locations, and the total amounts of energy that can be extracted is small compared with, for example, wind power. Still, at the endowed locations, very large hydropower plants can be established if correspondingly large elevated reservoirs can be formed, which depends on other possible uses of such areas.

The Fig. 2.20 shows the distribution of potential wind power extraction. Wind has been used for propulsion at sea long before turbines for

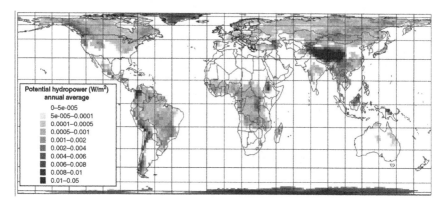

FIGURE 2.19 **Potential hydropower (W/m²) estimated from runoff at the Earth's surface (NOAA, 2013).** To calculate the capacity of a hydropower facility one would multiply by the catchments area (in m²). The water may reach the turbines after assembling in a reservoir lake or a water stream, or after traveling along the surface or as subsurface ground water. The estimate is based on the height difference between the surface receiving the precipitation and the sea level. In practice, the water may initially have travelled to a lower altitude by the mechanisms suggested earlier, and further, it may not be possible to establish the turbine water exit at sea level but only at some elevation.

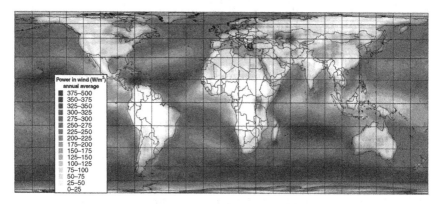

FIGURE 2.20 **Annual average power production from typical large wind turbines (hub height about 75 m, in W/m² for year 2000).** Collection would in practice be restricted to land areas and offshore areas with depths suited for foundation work, currently around 50 m (based on NCAR, 2006; Milliff et al., 2004, using the method proposed in Sørensen, 2008).

mechanical or electrical energy production were invented. A wind turbine cannot bring the wind to a standstill, so there is a limit to conversion efficiency (about 59%), but also a limit to how close wind turbines can be placed to each other without diminishing the winds received by turbines in the wake of others. A number of complex processes are available for restoring the wind speed by interaction with nearby air masses, on the sides and above. The turbine-spacing limit is roughly that the ratio of turbine swept area to the average land area available to each turbine is not larger than about 0.001, for land-based wind turbines. For ocean wind farms, the ratio may be a bit larger (Sørensen, 2010, 2015). Fig. 2.20 shows that the best wind conditions for power extraction is offshore or close to coasts, but presently, no proven technology is available for extracting wind power except by turbines mounted on foundations stretching to the bottom of the sea.[1] Thus, the current limit to extraction is that the water depth must be less than about 50 m.

Solar energy allows green plants to produce the biomass that we depend upon, for food (involving all sectors of farming: cereal crops, vegetables, fruit, and fodder for livestock), timber and a number of products derived from wood and plant materials. In terms of energy, the conversion of solar radiation by plants is of low efficiency (on average below 0.2%, rarely above 2%; Sørensen, 2010). This is the reason that large areas have had to be set aside for farming and forestry, a fact that opens the possibility of using biomass not suited for the primary purposes for energy tasks, as it has been done throughout human history and also before the introduction of agriculture (burning biomass for heating, cooking, and later many other tasks, such as metal production). Today, although burning wood is still common in the least developed regions of the world (plus regrettably for cosiness in some rich countries), it is discouraged in any advanced society due to the severe pollution caused by wood combustion in smaller burners not equipped with particle filters and other emission-reducing additions. A future use of secondary biomass for the production of liquid or gaseous biofuels is seen as a more acceptable pathway, at least for an interim period where especially the transportation sector has few alternatives.

The geographical distribution of net biomass production is shown in Fig. 2.21, for the case where no artificial irrigation is applied. Several areas currently using irrigation have been demonstrated to cause unsustainable use of surface water or to lower of the ground water level (Wada and Bierkens, 2014). Due to the large areas already committed to agriculture and forestry, the residues from these activities have the potential to deliver a sizeable part of the energy needs of societies, considering that biofuel production largely can avoid emissions and collect undesirable substances for responsible disposal or further transformation into useful materials (Sørensen, 2010). As with current uses of biomass, there is a fundamental requirement for returning nutrients to the soil in order to ensure long-term sustainability. Part of this task is carried out by livestock manure, when rotation between crop and range uses of agricultural land is practiced or by the farmer spreading collected manure, but also the abundant subsoil populations of decomposing agents form an essential

---

[1]Floating wind turbines are considered, but as yet unproven.

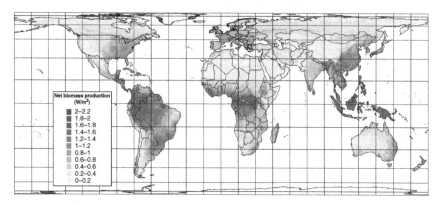

FIGURE 2.21 **Potential net annual biomass production (W/m²), whether natural vegetation or managed crops and forests (based on Melillo et al., 1993).** The production may be increased by irrigation, but this would be an unsustainable interference with the water cycle in many regions, if used on a truly large scale (Wada and Bierkens, 2014).

part of the functioning and sustainability of plant ecosystems (Bardgett and Putten, 2015). Any new technology for using biomass must honor the nutrient management requirement, but because it is a fact that current biomass users often neglect nutrient recycling, there is presently reservation expressed by ecologists against any additional use of biotechnologies.

2.1.4 Other Resources

The resources described in Section 2.1.2 were in the category of 'dead matter', while the biomass resources considered in Section 2.1.3 constitutes the plant part of what would then be called 'live matter', i.e., the organic components that plants and animals are made of. It would thus seem fitting to round this resource chapter up by describing the variety of resources associated with animals or humans.

In history, animals have played several roles, from being food prey, to dangerous threats to get rid of, occasionally becoming domesticated sources of companionship, serving as hunting partners, as agents of protection, or livestock. As a complement to vegetable food, animals represent a store of food energy. Earlier, when food preservation by freezing was not available, this could help to avoid starvation following a failed harvest. In any case, the solar energy conversion efficiency of the chain of processes leading to production of animal food is considerably lower than that of producing vegetable-based food. Recently, additional worries about biodiversity and loss of species have been raised, starting with mammals but eventually considering that other species should perhaps also receive attention (Scheffers et al., 2012).

Humans see themselves as having conquered the main stage of the Earth, with diminishing roles for 'wild' animals not being managed by humans. However, this is primarily true in relation to mammal species and other 'large' animals. Bacteria are far more abundant than humans, occupying not only the human gut but also an entire range of environments, from soils to plants and animals, and the same can be said of viruses. Other 'small' animals, such as soil nematodes are also very abundant, as are the somewhat larger insect animal species of ants, flies, and mosquitoes. Also birds seem at least as numerous as humans. Many of these animals are very well adapted to life on Earth and see humans as a convenient resource, for example, for food or as host for parasitic action. Species, such as ants have a higher resistance than humans to extreme conditions, for example radioactive fallout after a nuclear war, and may thus be the surviving inhabitants of the planet after the human species has become extinct. The late astronomer Carl Sagan and coworkers launched a project in 1980, involving search for intelligent life on other planets in the universe. A component in this program was to ask people to lend their computers to analyse signals from space when they were not in use by the owner. The surprising result that no intelligent messages were received has been interpreted as saying that when inhabitants of distant planets get so intelligent that they can send and receive messages to other solar systems, they also have invented nuclear weapons, and therefore likely destroyed their planets before they thought of communicating with us. In fairness it should be said that other interpretations of the missing results are possible: maybe intelligent beings realize that advanced inhabitants of other planets in the universe are most likely hostile, so the best strategy is to stay hidden by not sending any messages into space (Nørretranders, 1999).

For livestock there are detailed statistics, such as the abundance of cattle, pigs, and chicken shown in Figs 2.22, 2.23, and 2.24, while for other species and particularly the insect and bacterial varieties, no precise data are available. Fig. 2.25 shows the distribution of the current human population, for comparison. It is seen that none of the distributions are uniform, but that the distribution of cattle is roughly similar to that of man, except for South East Asia. In China, the abundance of pigs and chicken is considerably larger than that of cattle, and the Middle East region has a high abundance of chicken, but for cultural reasons not of pigs. Cattle are traditionally abundant where range areas suitable for grazing are present, and pigs on mixed-pasture land, but with increasing competition from stable-based cattle and pig farms where the animals are fed industrially produced fodder. The size of such farms influences disease frequency and possible spread to humans. Chicken may be raised in denser human settlements, such as those found, for instance, in eastern China. This has well-known implications for the spread of diseases that can be transferred from animals to humans, such as influenza varieties.

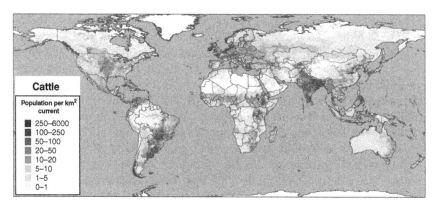

FIGURE 2.22   Current distribution of cattle (census data from FAO, 2015b; Wint and Robinson, 2007).

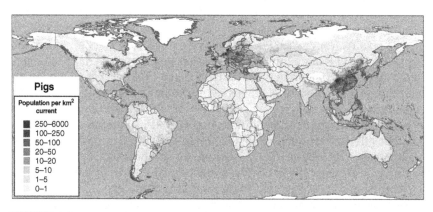

FIGURE 2.23   Current distribution of pig population (census data from FAO, 2015b).

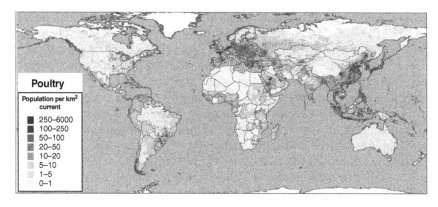

FIGURE 2.24   Current distribution of poultry population (census data from FAO, 2015b).

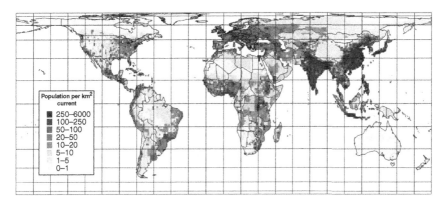

FIGURE 2.25   Current distribution of human population (UN, 2007).

Intellectual resources, such as art, literature, music, and skills based on education were mentioned in the property Section 2.1.1, noting that monetary valuation is mostly available for copyrighted material, and largely excluding items, for which the originator has been dead for more than the legal 50–70 years. The abundance of older works is variable. Literature is very abundant because it can be reprinted and digitally redistributed with no loss of utility, while digital renderings of artwork are usually not considered as cultural assets on the same footing as the original painting or sculpture. Musical compositions are like books regarded as cultural values for their contents, making availability of the original score manuscript less important and the same is true for scientific treatises describing important new insights. In terms of orders of magnitude, our literary inheritance may consist of several million works, our musical inheritance of some hundred thousand scores and the art inheritance at a similar level, depending on where the division line between art and applied-art-design is placed. More broadly, cultural resources comprise architecture (e.g., for buildings), historical and archeological sites, as well as the artefacts they contain, or which have been moved to museums. Both museums and cultural sites are abundant in most of the world (numbers run in tens of thousands), and accommodate cultural values ranging from those of central interest to humanity to others of more local interest. Fig. 2.26 shows the sites designated as World Heritage sites by UNESCO. They include both outstanding cultural sites and sites with exceptional natural features.

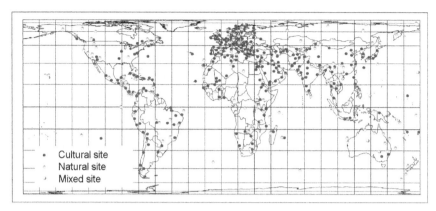

FIGURE 2.26   Indication of the location of UNESCO World Heritage sites (based on public lists, cf. UNESCO, 2015).

2.2 CURRENT PARADIGM OF ECONOMIC ORGANIZATION

2.2.1 The Liberal Market Economy

Trade and markets have played an important role since the early formation of human settlements. Evidence shows that hunting and gathering societies not only scavenged large areas for prey but that they travelled to neighboring villages more than hundred kilometres away to seek marriage, and that they were able to travel over 500 km to obtain special equipment or artwork, such as a particularly sharp hunting knife or a sea shell broche for a lady back in their village situated far from the ocean (Sørensen, 2012a).** Trade was by exchange of physical goods (hides, stone, and later metal tools), but also of items with a socially assigned and agreed value, such as amber or gold. Following the introduction of agriculture, trade in cereals became important, at first for the inauguration of farming in areas not previously engaged in this activity (notably the hunting societies) because they needed sowing seeds, as the plant varieties having grown successfully in the Middle East (wheat, barley) were not native to, for example, Europe. Later, trade of grain and livestock became common as a remedy for the year-to-year variations in successful farming for each

---

**Actually the notion of a stone age male buying the broche for his female companion is poorly substantiated, as there is at least some evidence that women took part in some of the trips and perhaps also in hunting. It could have been only trips associated with migrational relocation that involved whole families, but no evidence contradicts the possibility that women also took part in trade journeys, while grandparents attended to the children.

local society. The use of symbolic tokens of wealth, such as coins or other types of money enabled trade also in cases where the buyer did not have an obvious physical item to exchange with. Evidence shows that any kind of produced goods (ships, wagons, etc.) were traded as soon as one society had mastered or surpassed others in manufacture, and it is no surprise, that the 'industrial revolution' two- to three-centuries ago, characterized by increased use of steam power in commodity production, led to further emphasis on trade. This was the situation that prompted scholars to think about the theoretical framework for trade and finance.

Adam Smith (1776) formulated his market theory on the basis of three essential assumptions:

- The actors in the marketplace are roughly of the same magnitude and bargaining strength.
- The actors make rational decisions based on the commonly accepted economic theory.
- All actors have access to whatever information that is necessary to make rational decisions.

Whether these conditions may have been approximately fulfilled at the 18th century time of introducing the liberal market theory is difficult to ascertain but today, they are certainly not fulfilled, and yet a widespread belief in the liberal market theory has seemed to emerge, particularly over the last 50 years. The power equality of market actors is certainly not fulfilled in the present situation with multinational giant companies and the full range down to personal companies with one or a few employees. Furthermore, companies do everything they can to hide information from competitors, thereby preventing them from being able to make 'rational' decisions. Without rational decisions, the market cannot work as stipulated by Smith, and one must expect major flaws stemming from this irrationality. Examples include the recurrent financial crises and 'bubble economies' caused by banks incorrectly estimating the value of assets, such as property, and issuing ghost loans, which turn out not to have any collateral security once the bubble bursts and inflated prices (read: inability of the market to reach the correct prices) revert toward factual ones.

Current believers in the capitalistic market economy would claim that the classical market economy has been remedied by adding new theoretical ballast in what they call neoliberalism. It is rarely defined in a precise way but seems to postulate that the three nonfulfilled Adam Smith conditions may not make the market theory invalid because the market is seen as having an ephemeral 'intelligence' allowing it to make the correct decisions for any economic system, regardless of flawed composition.[††]

[††]A limited example is the claim that markets may function in violation of the enterprise similar-size requirement (Gand and Quiggin, 2003).

The essence of neoliberalism is thus to let the market (and preferably a global one encountering no trade barriers) govern all economic development with no interference from governments or individuals, demanding maximum deregulation and privatization. The only role seen for the state is to enforce property rights (DeMartino, 1999, 2000). Critics of the neoliberal market variant characterize it as trying to maximize inequality in society (Braedley and Luxton, 2010; Lavoie, 2014; Kotz, 2015). In this, neoliberalism has indeed been quite successful as demonstrated by Piketty (2014) and the study by World Bank (2011) discussed in Section 2.1.1. That the people in the lower part of the wealth scale are accepting such a fate and even casting their votes on neoliberal political parties is explained by the massive advertising campaigns of questionable content that have replaced serious popular political debate, but it is also due to a more subtle feature of neoliberalism, which is the worship of indiscriminate growth. This phenomenon will be further discussed in Chapter 3, but its role in hiding inequality leads to the concept that large economic growth can be combined with a slight improvement in wealth for the poorest strata of society (sometimes called the 'trickle-down effect') and still make inequality increase because although each new contribution to wealth is predominantly being collected by the already wealthiest layer, the autonomous market mechanisms (with a concern not just for profit but also for the related volume of consumption) make sure that the lowest strata in society are not experiencing or at least not noticing a fall in wealth. In reality division lines, such as the 'poverty limit' used in some countries are not static, and the real development over the recent 25 years may be argued to be one of increased poverty (Quiggin, 2009).

During the late 20th century, neoliberalism replaced the Keynesian view of the state as ultimately governing economic development by mending the flaws that kept appearing as the market failed to make the right decisions (as prompted by the 1929 financial crisis that started in the United States and subsequently spread globally). For some decades, the dominant economic policy in large parts of the world involved collaborative governance executed jointly by the government and the private sector. The 2007 financial collapse still being felt all over the world should have brought the neoliberal model to a fall, but no current governments seem to have the courage to dismiss their mainstream economic advisors, which tend to regard the crisis as a passing fluctuation. According to Quiggin (2010, 2013), the only thing that has recovered after the crisis is the self-esteem of mainstream economists.

2.2.2 The Substitution Hypothesis

Availability of raw materials is essential for any industry, and as indicated in Section 2.1.2, the abundance of mineral resources is highly variable and in several cases, proven resources will last only a few decades. It is therefore relevant to ask, how stable functioning of societies based as essentially as ours on manufactured goods can be ensured. The section about minerals indicates options for recycling, but shows that with current technologies, 100% recycling is not possible. The economic support for this assertion is that of the minerals we have mined and incorporated into various products, a part will always be released during manufacture, usage or end-of-life disposal to the environment (air or waterways) in a form so diluted that a secondary extraction from rivers and oceans or from air or land deposition following rainout will be orders of magnitude more expensive than the first extraction from the currently used geological deposits.

Since the total amount of resources on Earth is finite, this means that eventually, our societies will have diluted mineral resources to an extent causing a severe reduction in the options for industrial manufacture, as we know it. Some would argue that this situation is too far away to warrant concern, but that belief is linked to the prevailing economic discounting practice, where a positive interest rate makes anything happening 25 years into the future unimportant and anything happening 100 years into the future irrelevant. Seen in light of this planet having been occupied by humans for over a million years, the question of what we leave for future generations takes a more compelling form. We simply must ensure that important mineral resources are not diluted to an extent making them useless even with future advances in technological ability.

Current economic thinking deals with such issues in the following way: Of course, recycling is recommended whenever the cost of recycling is lower than that of new mineral extraction, which the market will ensure happens once the easily accessible geological resources are exhausted. However, the nature of the market-based theory is such that all minerals and other materials discarded before the price of new materials reaches that of recycled materials will not be available for recycling, at least in cases where the disposal of the materials leave them in less concentrated form than required for recycling (e.g., after incineration, today the most common way of handling refuse). Thus, only a small fraction of the original resources will be available to society at the time where the market economy wants to start employing recycling, namely the fraction residing in products at that moment. This is another example of the fallacy of the neoliberal mantra of 'let the market decide', but it is not the only way the current economic paradigm handles the raw material issue. The other line of defence is that of substitution.

Realizing that the incorrect handling of the timing for introducing recycling may lead to unavailability of minerals and other materials needed by industry, the suggestion of substitution is put forward: Just find another mineral that will do the same job. Usually, this does not always prevent short-term crises because it may take some time (shorter or longer) for the scientific and technological establishment to develop substitution options. The question is of course if the alternatives have been made ready by development prior to their necessity, and this is the basis for most societies actually spending money on research and development (R&D), despite the fact that the neoliberal economic paradigm would not allow money to be spend on such things before the market says it is necessary. Currently, the nations most enthused by neoliberal thinking are systematically removing funds from research and development, and although this is primary hitting public funding of such activities, the private sector enterprises following the prevailing economic paradigm are doing the same. For example, when several states in the United States privatized the electric utility system during the 1990s, R&D departments were closed and the following decades saw increasing numbers of blackouts in a supply and transmission system that became increasingly antiquated (OECD/IEA, 2005).

Concretely defined, the substitution hypothesis states that if a needed resource becomes depleted, another one will take its place, possibly at the higher price already prevailing and thus presumably accepted when the initial resource draws toward its end. This solution is market-driven and according to the neoliberal economists, there are no resources that cannot be substituted by new materials, and such new materials will always be developed when the marked says it needs them, with minimal delay.

Substitution has sometimes worked, other times not. An example of lacking preparation for substitution is provided by what has been called the 'first oil crisis' in 1973. The US oil producers wanted to raise prices so that extraction within the United States would again be feasible by use of the more expensive 'enhanced' extraction methods to get oil from deposits where as much had already been extracted as could be done with simple methods (Friberg et al., 1974). The Middle-East oil-producing countries at first declined raising prices, but after the Yom Kippur War they decided to both double prices and launch a supply embargo on Western countries. Since no one had developed alternatives to oil that could be implemented at short notice, and because no strategic oil stores had been built, the negative effect of the embargo and price hike became far more devastating for Western economies than it needed to have been. The new doubling of Middle-Eastern oil prices in 1980 did not have nearly the same impact on the Western economies, partly due to new oil fields having been developed outside the Middle East (North Sea, Mexican Gulf, accepting that oil production at much higher production costs could be profitable) and partly because newly established strategic stores ensured that supply to consumers were not immediately disrupted, while a number of alternative technologies (notably based on natural gas and coal) had already been put in place and were now ready to use.

An example of a successful substitution introduction is found in the development of silicon solar cells. Until the 1990s, the solar cell industries were using scrap material from the microelectronics industry as a raw material for silicon cells because it was inexpensive and available in amounts sufficient for the nascent solar cell demand. When demand increased, this was no longer the case. Microelectronic chips are increasingly miniaturized, while solar cells need to cover a large area to be intercepted by solar radiation and therefore cannot benefit from miniaturization, except for making the cells thinner, which has actually been achieved by replacing monocrystalline by multicrystalline technologies. On the other hand, solar cell silicon does not need to have the same level of purity and precision in doping (the process that adapts the material's nuclear properties to the frequency spectrum of solar radiation) as the microelectronics silicon material, which must allow very small patterns of electron pathways, and therefore the solar industry developed a 'solar-grade' silicon material that after a few years of learning became cheap enough to replace the microelectronics scrap material. The total solar cell price, that had declined systematically from the space cells of the 1970s and 1980s, stayed constant or slightly rising for a few years, but after 2005 again continued its decline and is now approaching a level where solar cells can viably enter energy supply systems in many locations with good access to solar radiation (Sørensen, 2012b). In this example, the substitution (involving two similar materials both of a lifetime exceeding 25 years) took place along with an ongoing development and thus avoided any disruption of supply.

Looking at the manufacturing industry in general, the last 50 year has seen massive substitution of organic polymer materials (plastic) where metals were used earlier. Even electrically conducting polymers have come into existence during the last decade. This makes it difficult to predict, which properties of each group of materials that may be essential and may refuse substitution. Medieval and Renaissance houses in northern Europe used to be made from burned clay bricks, having a bright red colour. This type of clay became totally exhausted by the mid-20th century, and for a while, houses had to be built from dull yellow bricks. Today, red bricks are back in the repertoire, due to synthetic dying, but a number of new building materials have been created and they sometimes offer possibilities not available earlier. An example is the façade elements with high structural strength, integrated heat insulation and no cool bridges that allow houses to be built with very low heat losses in winter and very low heat intrusion during summer. Before such materials were invented, there would be beams and poles providing the structural strength, and insulating material was put between the structural elements, unable to prevent heat to be conducted by way of the higher heat conductance of the structural beams. The bottom line is that it would be very difficult to say, if there are durable materials that fundamentally cannot be substituted.

While this is true for durable materials, it is not for nonrenewable material, such as fossil or nuclear fuels. When these are used they are, as mentioned previously, converted into substances too dilute to be re-used, and real depletion is thus possible. For nuclear fuels, this is a two-step process because current light-water reactors use nuclear fuels in a way that makes the reserves no larger than those of oil and natural gas, but where future use in breeder reactors could extend the effective period to depletion considerably (although still very short compared with timescales of human societies). The substitution hypothesis, which originally seemed to work well among fossil energy resources (Hubbert, 1962), under the assumption that the prices of different fuels would be similar once their conversion was suitably developed, is now appearing invalid because substitutions offered currently have much higher prices and then require that the energy sector assumes a dramatically larger share in the total economy, which is not inviting an economic business-as-usual future. One must also note that the substitution of plastic (organic polymers) for metals mentioned earlier is only a short-term solution, as the main raw material for plastic is fossil fuels and particularly oil. If other options can appear in the future is not known.

The verdict on the substitution hypothesis is vague. In the long-term, substitution of course will have its limits, while in the near future only resources that are indisputably depletable will require a detailed investigation of possible alternatives, their costs, and the time it will take to make them viable, technically and maybe economically. If economic viability is not possible, then arrangements of society need to be made that make the substitutes work anyway, with adaptation of the social organization to higher prices. Nearly all the critical remarks of a general character directed against infinite substitution options were formulated in a paper by Ehrlich (1989).

2.2.3 Externalities and Life Cycle Analysis

Market prices reflect perceived customer demands and products offered for sale. Externalities, such as environmental and social costs are only included if the customers actively demand to make additional payment for this. As seen from the side of the industry offering the products, there is a cost of manufacture and a minimum acceptable profit, combining to set a markup price. It is possible for the market price to be temporarily lower than the markup price, as part of competitive strategies, but in the long range, this is of course not acceptable to the manufacturer. Still, a low price induced by competition, say, due to too large overall capacity in the competing businesses (example: air travel) may over time have the effect of invoking increasing efficiency in manufacture or conduction of business, so that the marked-up cost may in this way get below the price desired by the market. This would be used as an argument for letting the market govern what happens, but unfortunately it is not the usual situation. The typical effect of the market-driven price setting is that externalities are neglected as much as it is possible, leaving the customers with products or services that appear cheap, but often are not when the indirect costs are considered.

The cost of environmental damage (pollution, climate change, land deterioration, habitat removal and lack of diversity) and social damage (health effects, poor work conditions, risk and accidents, change in human rights and political conditions) can in principle be evaluated and sometimes quantified in monetary terms by a lifecycle analysis and following assessment (Sørensen, 2011). Numerous examples show that the lifecycle cost, including costs occurring from the technology cradle to grave for both the product and the side-chains providing inputs and outputs to the product during its lifetime, can be quite substantial, and that restricting cost comparisons to direct cost can be very misleading because some technologies have few externalities, while others have total costs dominated by externalities. This means that the neoliberal focus on direct cost is generally false and will lead to wrong decisions made by consumers as well as by societies.

Advocating that the market will arrange all to the best is easy in a world where several previous decades have established a number of social safeguards that would not have emerged in the minimal government paradigm. Food items are sold in marketplaces, but with regulations ensuring they do not containing harmful substances, are not too old for consumption and based on farm products under continual agricultural authority inspection. Houses are built and sold in conformity with regulations, such as building codes and electricity norms, ensuring safety from collapse or unsafe installations. When the industry and commerce responsible for the products and services traded in society still chose to obey some rules of decency, it is because they see that not all consumers are of the neoliberal persuasion and they would like to retain those who are not as customers. To cater to that group, companies even find it useful to create a 'green' image for their enterprise. For instance, there is currently a choice of both energy-efficient cars and less efficient cars, and even some four-wheel drive offroad vehicles are becoming fairly efficient.

One would have to conclude that the current attitude toward the 'full costing' methods offered by life cycle analysis and more simplified versions of externality inclusion is mixed and that even those who generally endorse the neoliberal paradigm often make exceptions in the form of at least rudimentary concessions to environment and health.

2.3 CURRENT PARADIGM OF SOCIAL ORGANIZATION

2.3.1 Basic Human Rights

Lists of basic human rights were made at the time of the French Revolution and were to some extent incorporated in the constitutions of some countries, including the USA. Later, parts of the human rights list entered revised constitutions of several additional countries, especially in the regions of complex social organization that were formed in Europe and refined after World War II. While many countries in other parts of the world have not fully endorsed the principle of human rights, the role of these has entered national debate in a major way wherever it was allowed. A common basis for these discussions and actual changes of constitutions has been the Declaration of Human Rights (UN, 1948) that the winning parties of World War II proposed and a large majority agreed on at the United Nations, formed 1945 in an attempt to create a forum better suited to dealing with international conflicts and questions than the League of Nations created with a similar purpose after World War I.

The UN Declaration from 1948 contains a number of far-reaching statements that are reproduced below in full, with a few remarks added about the current status relative to the articles of the document, as well as remarks on where there may have arisen a need to expand or modify the text, although these are very few in number. The Declaration starts with a sharp and beautifully written preamble (in a fluent legal language!), which needs no updating:

> Whereas recognition of the inherent dignity and of the equal and inalienable rights of all members of the human family is the foundation of freedom, justice and peace in the world,
>
> Whereas disregard and contempt for human rights have resulted in barbarous acts, which have outraged the conscience of mankind, and the advent of a world, in which human beings shall enjoy freedom of speech and belief and freedom from fear and want has been proclaimed as the highest aspiration of the common people,
>
> Whereas it is essential, if man is not to be compelled to have recourse, as a last resort, to rebellion against tyranny and oppression, that human rights should be protected by the rule of law,
>
> Whereas it is essential to promote the development of friendly relations between nations,
>
> Whereas the peoples of the United Nations have in the Charter reaffirmed their faith in fundamental human rights, in the dignity and worth of the human person and in the equal rights of men and women and have determined to promote social progress and better standards of life in larger freedom,
>
> Whereas Member States have pledged themselves to achieve, in co-operation with the United Nations, the promotion of universal respect for and observance of human rights and fundamental freedoms,

Whereas a common understanding of these rights and freedoms is of the greatest importance for the full realization of this pledge,

Now, Therefore THE GENERAL ASSEMBLY proclaims THIS UNIVERSAL DECLARATION OF HUMAN RIGHTS as a common standard of achievement for all peoples and all nations, to the end that every individual and every organ of society, keeping this Declaration constantly in mind, shall strive by teaching and education to promote respect for these rights and freedoms and by progressive measures, national and international, to secure their universal and effective recognition and observance, both among the peoples of Member States themselves and among the peoples of territories under their jurisdiction.

Following this acknowledgement of what defines free human beings in a community setting, the declaration states in 30 articles the rules it sees as needed to make the human societies function:

1. All human beings are born free and equal in dignity and rights. They are endowed with reason and conscience and should act toward one another in a spirit of brotherhood.
2. Everyone is entitled to all the rights and freedoms set forth in this Declaration, without distinction of any kind, such as race, colour, sex, language, religion, political or other opinion, national or social origin, property, birth or other status. Furthermore, no distinction shall be made on the basis of the political, jurisdictional or international status of the country or territory, to which a person belongs, whether it be independent, trust, nonself-governing or under any other limitation of sovereignty.
3. Everyone has the right to life, liberty and security of person.
4. No one shall be held in slavery or servitude; slavery and the slave trade shall be prohibited in all their forms.
5. No one shall be subjected to torture or to cruel, inhuman or degrading treatment or punishment.
6. Everyone has the right to recognition everywhere as a person before the law.
7. All are equal before the law and are entitled without any discrimination to equal protection of the law. All are entitled to equal protection against any discrimination in violation of this Declaration and against any incitement to such discrimination.
8. Everyone has the right to an effective remedy by the competent national tribunals for acts violating the fundamental rights granted him by the constitution or by law.
9. No one shall be subjected to arbitrary arrest, detention or exile.
10. Everyone is entitled in full equality to a fair and public hearing by an independent and impartial tribunal, in the determination of his rights and obligations and of any criminal charge against him.
11.1. Everyone charged with a penal offence has the right to be presumed innocent until proved guilty according to law in a public trial, at which he has had all the guarantees necessary for his defence.
11.2. No one shall be held guilty of any penal offence on account of any act or omission, which did not constitute a penal offence, under national or international law, at the time when it was committed. Nor shall a heavier penalty be imposed than the one that was applicable at the time the penal offence was committed.

The first articles set out a number of basic human rights and the following ones give guidance for how to set up civil law and define admissible ways of conducting police and judicial work: forbidding arbitrary arrest, torture during detention and requiring guilt to be decided by courts in innocent-unless-proven-otherwise procedures conducted under full public scrutiny. These paragraphs are, for example, systematically violated by countries, such as the United States, in its processes against suspected terrorists (people disliked seem to currently be as routinely termed terrorists as they were termed communists 50 years ago). They are allegedly arrested and kept jailed in secret, without notifying anyone, are subjected to water boarding and other forms of "enhanced interrogation," and are not allowed a fair trial in an open court. Of course, many other countries currently violate human rights, but what makes the United States special is the hypocrisy of claiming that they interfere with the affairs of other countries in an effort to teach them democracy, freedom and justice! The following articles deal with rights of each citizen:

12. No one shall be subjected to arbitrary interference with his privacy, family, home or correspondence, nor to attacks upon his honour and reputation. Everyone has the right to the protection of the law against such interference or attacks.

13.1. Everyone has the right to freedom of movement and residence within the borders of each state.

13.2. Everyone has the right to leave any country, including his own, and to return to his country.

14.1. Everyone has the right to seek and to enjoy in other countries asylum from persecution.

14.2. This right may not be invoked in the case of prosecutions genuinely arising from nonpolitical crimes or from acts contrary to the purposes and principles of the United Nations.

15.1. Everyone has the right to a nationality.

15.2. No one shall be arbitrarily deprived of his nationality nor denied the right to change his nationality.

16.1. Men and women of full age, without any limitation due to race, nationality or religion, have the right to marry and to found a family. They are entitled to equal rights as to marriage, during marriage and at its dissolution.

16.2. Marriage shall be entered into only with the free and full consent of the intending spouses.

16.3. The family is the natural and fundamental group unit of society and is entitled to protection by society and the State.

17.1. Everyone has the right to own property alone as well as in association with others.

17.2. No one shall be arbitrarily deprived of his property.

18. Everyone has the right to freedom of thought, conscience and religion; this right includes freedom to change his religion or belief, and freedom, either alone or in community with others and in public or private, to manifest his religion or belief in teaching, practice, worship and observance.

19. Everyone has the right to freedom of opinion and expression; this right includes freedom to hold opinions without interference and to seek, receive and impart information and ideas through any media and regardless of frontiers.

20.1. Everyone has the right to freedom of peaceful assembly and association.

20.2. No one may be compelled to belong to an association.

Article 12 forbids interference with the personal life of human being, including the collection of information from letters (and Internet visits) and other abuses of privacy, such as the US National Security Agency's spy activities on global phone conversations, emails and Internet surf-patterns. The following articles ensure freedom of movement, but accept that the world is divided into country entities that may be visited (without political persecution) but where permission is needed for a foreigner to take permanent residence. The strong emphasis on nations is appropriate today, but other forms of global organization with fewer rights accorded to national entities are possible, as will be further discussed in Chapters 4 and 5.

Article 16 deals with the freedom of marriage. Some would disagree with paragraph 16.3 stating that the traditional family is the basis of society, and problems of overpopulation may imply the necessity to restrict the number of children that a woman can have, as it has been tried in China. Also the provisions regarding rights to own property could be seen as based on a particular economic arrangement, which is not of a sufficiently universal validity to warrant inclusion in the declaration of human rights (cf. the discussion in Section 2.1.1 of rental arrangements substituting ownership for some types of property, namely those with infinite value, as further elaborated in Section 3.1).

Articles 18 and 19 contain fundamental statements of the rights of freethinking, of holding any beliefs (e.g., of a religious nature) and free expression, whether in speech or otherwise. These are personal rights although it is stated that they could be exercised in unison by a group of people. Free expression includes free flow of information in any type of media across national borders. The nature of rights as pertaining only to individual citizens is emphasized in Article 20, stating that individuals may but cannot be forced to belong to any association, whether political, religious or of any other kind. Specifically, no rights are given to institutions. This article is today de facto violated not only by Islamic extremists, but in many cases by all Islamic institutions, just as they have been violated by the Roman Catholic Christian Church through its early cleansing campaigns, its crusades and inquisition, as well as its forced conversion to Christianity during the colonial period.

21.1. Everyone has the right to take part in the government of his country, directly or through freely chosen representatives.

21.2. Everyone has the right of equal access to public service in his country.

21.3. The will of the people shall be the basis of the authority of government; this will shall be expressed in periodic and genuine elections, which shall be by universal and equal suffrage and shall be held by secret vote or by equivalent free voting procedures.

22. Everyone, as a member of society, has the right to social security and is entitled to realization, through national effort and international co-operation and in accordance with the organization and resources of each State, of the economic, social and cultural rights indispensable for his dignity and the free development of his personality.

Article 21 about governance is remarkable as it does not take a stand on whether democracy should be direct or through representatives. Article 22 is a sweeping plea for everyone's right to be accorded secure economic, social and cultural conditions. Mentioning that the agents securing these rights can be national or international is a statement on the role that the United Nation hoped to be increasingly accorded, extending the modest start during the early post–World War II years.

23.1. Everyone has the right to work, to free choice of employment, to just and favorable conditions of work and to protection against unemployment.

23.2. Everyone, without any discrimination, has the right to equal pay for equal work.

23.3. Everyone who works has the right to just and favorable remuneration ensuring for himself and his family an existence worthy of human dignity, and supplemented, if necessary, by other means of social protection.

23.4. Everyone has the right to form and to join trade unions for the protection of his interests.

24. Everyone has the right to rest and leisure, including reasonable limitation of working hours and periodic holidays with pay.

Articles 23 and 24 sets out the rights to work and leisure time, with remuneration, but may be criticized for being vague on the possibility of work arrangements different from those of current salaried workers and employees serving enterprise owners. Furthermore, the right to form worker's unions would seem already covered by Article 20 and it is not clear why such unions should have a fundamental role in society, particularly because their methods of operation and providing help to workers is not defined in the declaration text. Here the United Nation is probably responding to problems characterizing the previous history of capitalism, but in a form not quite possessing the universality aimed at in the declaration.

25.1. Everyone has the right to a standard of living adequate for the health and well-being of himself and of his family, including food, clothing, housing and medical care and necessary social services, and the right to security in the event of unemployment, sickness, disability, widowhood, old age or other lack of livelihood in circumstances beyond his control.

25.2. Motherhood and childhood are entitled to special care and assistance. All children, whether born in or out of wedlock, shall enjoy the same social protection.

26.1. Everyone has the right to education. Education shall be free, at least in the elementary and fundamental stages. Elementary education shall be compulsory. Technical and professional education shall be made generally available and higher education shall be equally accessible to all on the basis of merit.

26.2. Education shall be directed to the full development of the human personality and to the strengthening of respect for human rights and fundamental freedoms. It shall promote understanding, tolerance and friendship among all nations, racial or religious groups, and shall further the activities of the United Nations for the maintenance of peace.

26.3. Parents have a prior right to choose the kind of education that shall be given to their children.

Article 25.1 starts spelling out the most fundamental rights of food and shelter, as well as the right to assistance in cases of health problem, unemployment, and other disabilities outside the individual's control. It is not specified who should provide the required services and pay for them. Moreover, the formulation 'himself and his family' indicates an unfortunate endorsement of patriarchal societies, which is not weakened by the fact that the other part of the article deals with motherhood. Article 26 states the basic right to education, careful not to restrict it to childhood schooling and emphasizing the important role of education in creating the conditions for securing respect for the human rights. Many will find it unfortunate, that paragraph 26.3 accords parents the right to decide what education their children should have. This may seem in the real world to be a prescription for avoiding progress in the direction of making the human rights generally accepted, and the possibility that governments could be more progressive than parents is excluded by according the education selection right to all adult citizens, independent of their own education. Even if unlimited democracy is generally in place, Article 26.3 would seem to allow any minority to educate their children in ways disregarding some or all of the human rights in the declaration. One may also note that gender issues are not highlighted in the declaration except in the preamble, but they can be said to have been dealt with by the general rights accorded to everyone.

27.1. Everyone has the right freely to participate in the cultural life of the community, to enjoy the arts and to share in scientific advancement and its benefits.

27.2. Everyone has the right to the protection of the moral and material interests resulting from any scientific, literary or artistic production, of which he is the author.

28. Everyone is entitled to a social and international order, in which the rights and freedoms set forth in this Declaration can be fully realized.

The cultural and scientific sides of activities taking place in a society are dealt with in Article 27, inviting everyone to enjoy but at the same time emphasizing the right to institute patent and copyright protection. Current implementations of these 'rights' are only modestly benefiting those creating the artwork, music or inventions of a scientific or technological nature, relative to publishers and companies reaping the profits for periods that by far exceeds the life of the originator (e.g., in Europe, copyrights currently extend 70 years after the death of the originator). Whether occasional inheritance of rights by children (usually not involved in the creation) is a remedy remains open to debate.

29.1. Everyone has duties to the community, in which alone the free and full development of his personality is possible.

29.2. In the exercise of his rights and freedoms, everyone shall be subject only to such limitations as are determined by law solely for the purpose of securing due recognition and respect for the rights and freedoms of others and of meeting the just requirements of morality, public order and the general welfare in a democratic society.

29.3. These rights and freedoms may in no case be exercised contrary to the purposes and principles of the United Nations.

30. Nothing in this Declaration may be interpreted as implying for any State, group or person any right to engage in any activity or to perform any act aimed at the destruction of any of the rights and freedoms set forth herein.

Article 29.1 is remarkable as it introduces duties that are connected with the benefits of human rights. This subject may be expanded, but its introduction here is farsighted. Articles 29 and 30 sum up the perseverance of the rights formulated in the declaration. A declaration of human rights should be universal, independent of time and place and as this brief review of the Human Rights Declaration seen through a 2015 eye shows, the 1948 UN documents come close to achieving this. Despite the abundant violation of these rights that takes place, they have indeed become part of the luggage that a large part of the world's population carries, if not always conscientiously.

Some countries have incorporated a number of the human rights into their constitutions, along with paragraphs on governance and occasionally additional topics. The formulation or updating of constitutions to make sure that they address both basic rights and pressing issues currently identified will be addressed in Section 3.2.

2.3.2 Governance and Political Setup

The map of various forms of government in function across the present world comprises a varied kaleidoscope of democratic and less democratic types of governance. The traditional Western countries all claim democracy, hold elections at intervals and teach children in school that all is under the best control, while many other countries have more or less dictatorial rule, even if they stage elections from time to time. Democracy is clearly more than having elections every four years. It includes respect for sizeable minorities, for human rights and for maintaining a sustainable relationship with the environment, whether local or global.

Current implementations of democracy are basically not using the classical Greek idea of direct democracy (Section 3.2.1), but rely on a representative democracy, which indeed was also the outcome of the attempts to shape a democracy in ancient Greece. The standard objection to direct democracy is that if people are asked to vote yes or no to paying taxes, a majority voting 'no' is likely, while the same voters probably would vote 'yes' to expenditures seen as benefiting them. This behavior would invalidate many of the functions usually expected from national entities, but the example could as well be taken to illustrate that democracy cannot be based upon uninformed voters, and that earning the right to vote should rather be tied to having proven that the issues voted upon are understood (more on this line of thinking in Chapter 3). However, in a representative democracy there may well be singular issues that are best determined by a general referendum, and most current democracies have provisions for staging such referenda, even if the basic governance is through representative parliamentary assemblies and executive governments.

The basis for current democratic implementations is one or more parliamentary congregations (one is becoming the most common preference) with members elected by a legally determined procedure. They constitute what is termed the legislative branch of government (issuing specific laws on any relevant matter), and are supplemented by an executive branch (ministers and their departments) and by a legal branch (courts to resolve disputes and ambiguities of interpreting laws). Some countries have a government headed by a prime minister, but at the same time a king or a president. Kings are of course left over from the time before democracy, and because they are hereditary and not elected, they should not have any power in a democracy. Yet constitutions are often unclear on this point or even accord a certain power to the king. The same is true for presidents, although they are usually elected and their power therefore more legitimate, but the relative distribution of power between presidents and parliament, and the roles of ministers (sometimes called secretaries) in government departments (including the prime minister usually heading a state department) as demonstrated by their choice in dividing loyalty between president and parliament, also differ between nations.

Procedures for election vary substantially between countries, but the tendency has been toward replacing person election by party election, meaning that one votes for a party and optionally state, which candidate in this party is preferred. The advantage of this scheme can be that if the person voted for already has enough votes to get elected or has far too few, then your vote may go to another person in the same party, possibly with similar opinions. However, the details of voting legislation may prevent such a vote transfer. Some countries have 'winner-takes-all' elections on both state and regional levels, where votes on other candidates than the winners are lost. If there are many candidates for an important position, say, for the election of a president, some countries have legislation requiring additional elections if no candidate gets over 50% of the votes, omitting in the second round the candidates with fewer votes than the two highest ones. For election of the members of parliament, it is often considered important that each region of a country is represented, and some countries have a fixed number of delegates to be elected in each region. This may also cause bias because the distribution of population originally used to set the number of delegates can change with time. For example, in the United Kingdom, such numbers of delegates were kept fixed from before the industrial revolution to recent times, causing cities with many workers to be grossly underrepresented in parliament, relative to smaller communities often dominated by people with employer outlooks. The same is true in the United States, where each state has a number of representatives based on population distributions valid some time ago. As a result, the presidential candidate getting most popular votes may lose the election (as it happened for Al Gore vs. George Bush Junior in the year 2000; Katz, 2001a). This is in violation of the very basic democratic principle that each vote cast should count the same. Some European countries have remedied this by a procedure invented by d'Hondt in Holland, in which some fraction of the parliamentary delegates are elected according to local preferences, while the rest are elected from otherwise lost votes on a national scale, by attributing same party votes until the average number of votes necessary for a seat is reached. Obviously, this procedure only works within a party-based frame of elections. Countries based on party democracy usually have a lower limit for parties (typically 2–5% of all votes), under which no seats in parliament may be won by the d'Hondt redistribution method. This means that 'lost votes' have not fully been eliminated. In any case it is a questionable presumption that if you vote for a person that is not elected, then you do not mind that your vote is donated to someone else in the same party.

Following parliamentary elections, a government is formed by adjusting department portfolios and selecting ministers for each department. The procedure for this also varies between countries. Some let the largest party try to form a government, then if it fails the next largest and so on. Others have the king or president decide on how the government establishment is accomplished. In the end, the goal is to establish a government, eventually as a coalition between more than one party that does not have a parliamentary majority against it (at least not initially). This may even be a minority government, if some parties do not want to be part of the government but also do not wish to overthrow it.

Reducing the political debates and deliberations to party quarrels obviously involves severe dangers because it largely restricts the area of political action to inflexible party programs and agreed policies, losing the pluralism of personal attitudes held by the totality of members of a parliament and even more those held by their constituencies. Some countries have constitutions stating that members of parliament, once elected, are bound only by their own conscience, but this cannot be so if the parliament is divided into a number of parties, requiring their representatives to vote in unison on each important issue. The situation becomes cloudier in the present political climate, where many parties seek election by making populist statements aimed only to please the voters and by paying for advertising campaigns (often conducted by professional media-manipulating companies). In both cases, there is no guarantee that the policy pursued will resemble the preelection statements and promises or that these are at all amenable to execution.

Mair (2013) flatly concludes that party democracy is no longer a valid or acceptable form of democracy. He explains the increasing indifference of citizens towards politicians and politics by noting that voters do not see lifetime party politicians as actually ruling the country: often groups outside the political parties decide important issues in defiance of democracy and unchallenged get away with influencing decisions by acting as 'stakeholders' (USA) or 'lobbyists' (EU), often in consort with civil servants in government departments, violating the professional detachment expected from them (Blinder, 1997; Katz, 2001b; Zakaria, 1997). Actually, requiring unbiased work from civil servants is a European specialty because European civil servants are not replaced when the government changes colours, in contrast to, for example, the USA, at least for the upper echelons of bureaucrats. The problem is not just with illegitimate influence on decisions by nonelected actors, but also the party-based democracy is accused of often acting for party promotion rather than for the constituency (Katz and Mair, 2009), just as politicians have been accused of promoting their own livelihood as lifetime social parasites.

Most of the near 50% of the current world not embracing the notion of democracy have governments and parliaments that often borrow several traits from their democratic counterparts. Some countries allow only one party but still stage elections and have political debates, although often hidden in the corridors of an oligarchic minority, but they are occasionally influenced by expressions voiced in the general public or by important enterprises. Other countries have an autocratic ruler not admitting public debate, but occasionally still being influenced, for example, by advisory bodies.

2.3.3 Welfare as Currently Viewed

Currently employed measures of welfare are rarely looking at wealth and at the relation between the two concepts defined in Section 2.1.1, but only at some annual flow rates, such as measures of economic activity. Of these, the one seen most often in statistical surveys is the gross national product, GNP. It is defined as the sum of all economic transactions, such as selling/buying goods and services in a year by the citizens of a nation. Often, only registered monetary transactions get incorporated into the GNP, not household work or other unpaid work or service provision. Regarding produced goods, care needs to be taken not to double-count market payments for exchange of raw materials, parts production, and the final goods. Transactions involving other nations are included in the GNP, in contrast to the concept of 'gross domestic product' (GDP) including only transactions within a nation. Less important subtleties involve, for example, the treatment of foreign taxation. The term 'gross' indicates that no deductions are made, for example, to account for depreciation. As regards services, it can be even more difficult to avoid double-counting and other irrationality. An example of transactions included in the GNP is the frequent currency exchanges that may be carried out, for example, by banks in an effort to earn from successfully forecasting changes in exchange rates. This could cause large increases in GNP without any tangible money being involved and the profit may again disappear over periods of a few hours. As another example, if my neighbor and I decide and register that she hands me a million dollars on even days, and I hand her a million on odd days, then the GNP will increase by 366 million dollars in a leap-year. Such a setup has changed nothing by the end of the year, which supports the view of the GNP as a rather useless artefact.

Different efforts to quantify welfare have been made (Section 2.1.1), for example by resorting to interview studies when data have not been available. Clearly, a sober measure of welfare would be much more useful than the GNP, and efforts to construct alternatives to GNP that would approach this goal will be presented in Section 3.1.2.

It should be mentioned that focus on welfare has led to practical results in the form of governance schemes giving welfare a central role. This is the case for the Scandinavian Welfare Economic Model that was implemented in the 1930s and onward in the three Scandinavian countries. It is based on

1. Free medical assistance (practicing doctors and hospitals);
2. Free education (from primary school to university level);
3. Pension schemes for everyone (free basic pension plus insurance type arrangement with running payments from salary or other income); and
4. Unemployment and reduced-ability-to-work compensation (basic diets plus voluntary insurance type arrangement).

In addition to these social benefits, the welfare economic model made use of the substantial public sector created to handle such payments, financed by progressive taxation to also create strategic sectors handling transport, energy production, mail and telecommunication, arguing that these services could not be left to the fluctuating market choices. However, all other sectors were left in private hands and no general nationalization was foreseen. This defined the Social Democratic political paradigm, as distinct from the socialistic one. The late 20th century transition to a neoliberalistic economic paradigm has so far not affected the four basic benefits listed, although debates on 'user-payment' have occasionally surfaced, but the paradigm change has led to privatization of nearly all the former strategic sectors.

References

Bardgett, R., Putten, W. van der, 2015. Belowground biodiversity and ecosystem functioning. Nature 515, 505–511.

Baturin, G., 1997. Mineral resources of the ocean. Lithol. Miner. Resour. 35, 399–424.

Blinder, A., 1997. Is government too political? Foreign Aff. (11/12), 115–126.

Braedley, S., Luxton, M. (Eds.), 2010. Neoliberalism and everyday life. McGill-Queen's University Press, Montreal.

Dasgupta, P., 2001. Human well-being and the natural environment. Oxford University Press, Oxford.

Davies, J., Sandström, S., Shorrocks, A., Wolff, E., 2008. The world distribution of household wealth. Discussion paper 2008/03, UNU-World Inst. Dev. Econ. Research, Helsinki.

Davies, J., Sandström, S., Shorrocks, A., Wolff, E., 2009. The global pattern of household wealth. J. Int. Dev. 21, 1111–1124.

DeMartino, G., 1999. Global neoliberalism, policy autonomy, and international competitive dynamics. J. Econ. Issues 33 (2), 343–349.

DeMartino, G., 2000. Global Economy, Global Justice: Theoretical Objections and Policy Alternatives to Neoliberalism. Routledge, Cambridge.

ECMWF, 2015. ERA-40 Reanalysis Project (Description in Kållberg et al., 2007). European Centre for Medium-Range Weather Forecasting, Reading. Available from Apps.ecmwf.int/datasets/era40-daily/ (assessed 2008.)

Ehrlich, P., 1989. The limits to substitution: meta-resource depletion and a new economic-ecological paradigm. Ecol. Econ. 1, 9–16.

Estes, R., 1984. The Social Progress of Nations. Praeger Publishers, New York.

FAO, 2015a. Fish species distributions. Food and Agriculture Organization of the United Nations, Fisheries and Aquaculture Department. Available from www.fao.org/figis/geoserver/factsheets/species.html

FAO, 2015b. Observed livestock densities. Food and Agriculture Organization of the United Nations (Description in Wint and Robinson, 2007). Available from www.fao.org/ag/againfo/resources/en/glw/GLW_dens.html

Fischer, I., 1906. Nature of capital and income. McMillan, New York.

Friberg, M., Hettne, B., Leksell, I., Tompuri, G., 1974. Utvikling, magt och energi i ett internationalt konfliktperspektiv. In: Jungen, B. (Ed.), Energi – inta endast en fråga om teknik. Swedish Parliamentary Hearing. Centrum för Tvärvetenskap, Gothenburg University, Gothenburg, Sweden, pp. 51–59 (in Swedish).

Gand, J., Quiggin, J., 2003. A technological and organizational explanation for the size distribution of firms. Small Bus. Econ. 21 (3), 243–256.

Goldewijk, K., Beusen, A., Janssen, P., 2010. Long-term dynamic modelling of global population and built-up area in a spatially explicit way: HYDE 3.1. Holocene 20 (4), 565–573.

Goldewijk, K., Beusen, A., Drecht, G., Vos, M., 2011. The HYDE 3.1 spatially explicit database of human-induced global land-use change over the past 12,000 years. Glob. Ecol. Biogeogr. 20, 73–86.

Graedel, T., Allwood, J., Birat, J.P., Buchert, M., Hagelüken, C., Reck, B., Sibley, S., Sonnemann, G., 2011. What do we know about metal recycling rates? J. Ind. Ecol. 15 (3), 355–366.

Hamilton, K., Clemens, M., 1999. Genuine savings rates in developing countries. World Bank Econ. Rev. 13 (2), 333–356.

Hamilton, K., Liu, G., 2014. Human capital, tangible wealth, and the intangible capital residual. Oxford Rev. Econ. Policy 20 (1), 70–91.

Hubbert, M., 1962. Energy Resources. Report to the Committee of Natural Resources, US National Academy of Science, Publication # 1000D.

Kållberg, P., Simmons, A., Uppala, S., Fuentes, M., 2007. The ERA-40 archive. ERA-40 Project Report Series No. 17, ECMWF, Reading.

Kaplan, J., Krumhardt, K., Ellis, E., Ruddiman, W., Lemmen, C., Goldewijk, K., 2010. Holocene carbon emissions as a result of anthropogenic land cover change. The Holocene 21 (5), 775–791.

Katz, R., 2001a. The 2000 presidential election: a perverse outcome? Representation 38 (2), 141–149.

Katz, R., 2001b. Models of democracy – elite attitudes and the democratic deficit in the European Union. EU Polit. 2 (1), 53–79.

Katz, R., Mair, P., 2009. The cartel party thesis: a restatement. Perspect. Polit. 7 (4), 753–766.

Kotz, D., 2015. The Rise and Fall of Neoliberal Capitalism. Harvard University Press, Boston.

Lavoie, M., 2014. Financialization, neoliberalism, and securitization. J. Post Keynesian Econ. 35 (2), 215–233.

Legatum Institute, 2011. Legatum prosperity index report. London 2010 report. Available from www.prosperity.com

Liu, G., 2011. Measuring the Stock of Human Capital for Comparative Analysis, OECD Statistics Working Paper No. 2011/06, doi:10.1787/5kg3h0jnn9r5-en.

Loveland, T., Reed, B., Brown, J., Ohlen, D., Zhu, J., Yang, L., Merchant, J., 2000. Development of a global land cover characteristics database and IGBP DISCover from 1-km AVHRR data. Int. J. Remote Sens. 21 (6/7), 1303–1330.

Mair, P., 2013. Ruling the Void – The Hollowing of Western Democracy. Verso, London.

McDonald, N., Pearce, J., 2010. Producer responsibility and recycling solar photovoltaic modules. Energ. Policy 38, 7041–7047.

McEvedy, C., Jones, R., 1979. Atlas of World Population History. Penguin Books/Allen Lane, London. Undated US edition by Facts on File, New York.

Megrey, B., Moksness, E. (Eds.), 2009. Computers in Fisheries Research. Springer, USA.

Melillo, J., McGuire, A., Kicklighter, A., Moore, B., Vorosmarty, C., Schloss, A., 1993. Global climate change and terrestrial net primary production. Nature 363, 234–240.

Milliff, R., Morzel, J., Chelton, D., Freilich, M., 2004. Wind stress curl and wind stress divergence biases from rain effects on QSCAT surface wind retrievals. J. Atmos. Ocean Tech. 21, 1216–1231.

Mohr, S., Mudd, G., Giurco, D., 2012. Lithium resources and production: critical assessment and global projections. Minerals 2, 65–84.

NCAR, 2006. Blended winds from the QSCAT/NCEP satellite mission (Description in Milliff et al., 2004). National Center for Atmospheric Research data site rda.ucar.edu/datasets/ds744.4

NOAA, 2013. NCEP-NCAR CDAS-1 monthly diagnostic surface runoff data. Available from the Columbia University data collection: iridl.ldeo.Columbia.edu/ sources/.noaa/ncep-ncar/cdas-1/

Nørretranders, T., 1999. Frem i tiden (Situations seen from cosmos). Tiderne Skifter, Copenhagen (in Danish).

Odell, P., 1997. The exploitation of off-shore mineral resources. Geojournal 42, 17–26.

OECD/IEA, 2005. Learning from the Blackouts. Energy Market Experience Series. Int. Energy Agency, Paris. www.iea.org/publications/freepublications/publication/Blackouts.pdf (accessed 2014.)

OECD, 2012. Inequality. In: Perspectives on Global Development 2012 (Chapter 4). Available from dx.doi.org/10.1787/persp_glob_dev-2012-en

Ohlsson, H., Roine, J., Waldenström, D., 2006. Long-run changes in the concentration of wealth. Research paper 2006/103, UNU-World Inst. Dev. Econ. Research, Helsinki.

Piketty, T., 2014. Capital in the twenty-first century. Harvard University Press, Boston.

Piketty, T., Saez, E., 2014. Inequality in the long run. Science 344, 838–843.

Ponisio, L., M'Gonigle, L., Mace, K., Palomino, J., Valpine, P., Kremen, C., 2015. Diversification practices reduce organic to conventional yield gap. Proc. R. Soc. B 282, 20141396.

Proudhon, P., 1840. Qu'est-ce que la propriété? (What is property?). La Librairie Prévot, Paris. Available from Project Gutenberg, www.gutenberg.org, e-book #360 (updated 2013).

Quiggin, J., 2009. Six refuted doctrines. Econ. Papers 28, 239–248.

Quiggin, J., 2010. Beauty ≠ truth? Thoughts on Krugman's 'how did economists get it so wrong?'. Agenda 17 (1), 113–118.

Quiggin, J., 2012. Equity between overlapping generations. J. Public Econ. Theory 14 (2), 273–283.

Quiggin, J., 2013. State of economics in 2012: complacency amid crisis. Econ. Rec. 89, 23–30.

Scheffers, B., Joppa, L., Pimm, S., Laurance, W., 2012. What we know and don't know about Earth's missing biodiversity. Trends Ecol. Evolut. 27 (9), 501–510, Erratum: 27 (12), 712–713.

Smith, A., 1776. An inquiry into the nature and causes of the wealth of nations. Methuen & Co., London, 1904. Available from www.econlib.org/library/Smith/smWN.html.

Sørensen, B., 2008. A new method for estimating off-shore wind potentials. Int. J. Green Energy 5, 139–147.

Sørensen, B., 2010. Renewable Energy – physics, engineering, environmental impacts, economics & planning, fourth ed. Academic Press, Elsevier, Burlington.

Sørensen, B., 2011. Life-cycle analysis of energy systems – from methodology to applications. RSC Publishing, Cambridge.

Sørensen, B., 2012a. A history of energy – Northern Europe from the Stone Age to the present day. Earthscan/Routledge, Cambridge.

Sørensen, B., 2012b. Hydrogen and Fuel Cells – Emerging technologies and applications, second ed. Academic Press/Elsevier, Oxford and Burlington.

Sørensen, B., 2015. Energy Intermittency. CRC Press/Taylor & Francis, Baton Rouge.

Tuncuk, A., Stazi, V., Yazici, E., Deveci, H., 2012. Aqueous metal recovery techniques from e-scrap: hydrometallurgy in recycling. Minerals Eng. 25, 28–37.

UN, 1948. The Universal Declaration of Human Rights. United Nations, www.un.org/en/documents/udhr/

UN, 2007. Populations 1996, 2015, 2050; World population prospects. United Nations Population Division, with update 2010, and UN Development Program. www.undp.org/popin/wdtrends/pop/fpop.htm and esa.un.org/unpp/.

UNEP, 2010. Metal Stocks in Society. Status Report 1 from the International Resource Panel of the United Nations Environmental Programme.

UNEP, 2011. Recycling rates of metals. Status Report 2 from the International Resource Panel of the United Nations Environmental Programme.

UNESCO, 2015. World Heritage Convention. United Nations Educational, Scientific and Cultural Organization. whc.unesco.org/en/interactive-map/

University of Columbia, 1997–2015. USGS EROS Landcover GLCCDB v2.0. Available from iridl.ldeo.columbia.edu/SOURCES/.USGS/.EROS/.LandCover/.GLCCDB/.V2p0/

USGS, 2015a. Global land cover characteristics database v2.0. United States Geological Survey. Available from edc2.usgs.gov/glcc/glcc.php

USGS, 2015b. Mineral Resources Data System. United States Geological Survey. Available from mrdata.usgs.gov/mrds.

Wada, Y., Bierkens, F., 2014. Sustainability of global water use: past reconstruction and future projections. Environ. Res. Lett. 9 (10), 104003.

Wint, W., Robinson, T., 2007. Gridded livestock of the world. FAO, Rome.

Woodhouse, A. (Ed.), 1951. Puritanism and Liberty, being the Army Debates (1647–1649) from the Clarke Manuscripts with Supplementary Documents. University of Chicago Press, available at http://oll.libertyfund.org/title/2183.

World Bank, 2006. Where is the wealth of nations? Washington DC.

World Bank, 2011. The changing wealth of nations. Available as e-book from http://data.worldbank.org/news/the-changing-wealth-of-nations, with associated spreadsheet data at data.worldbank.org/data-catalog/wealth-of-nations

Zakaria, F., 1997. The rise of illiberal democracy. Foreign Aff. (11/12), 22–43.

Zhu, X., 2014. GIS and urban mining. Resources 3, 235–247.

3

New Paradigms: In Search of a Third Way

3.1 NEW INDICATORS OF SOCIAL WELFARE

3.1.1 Growth, but of What?

Growth is imbedded in all human beings as a positive concept from the day we are born. Once we were all babies but wanted to grow to become children with a larger action radius. As children we wanted to grow to become adults, experience love and sex, and to form our own living quarters reflecting our personal style. It has therefore been simple for human beings to accept the worship of growth presently preached by economists and politicians. But the problem is that the term 'growth' is attributed different meanings in different circumstances. Some of these may conform to the childhood notion of growth being desirable, but others cover more specialized types of growth that we should not accept without critical questions. Foremost among these is the economic growth defined as growth in the very artificial quantity called the gross national product (GNP).

The GNP, as described in Section 2.3.3, has primarily appeared as a convenient indicator because it is very easy to extract from the kind of data that national statistical bureaus have specialized themselves in collecting. It measures economic activity and includes anything that involves the exchange of money. No questions are asked regarding whether the activity is useful, whether it brings benefits to society or to individuals, or whether it has negative impacts on society or environment. Given this background, it is strange that political leaders have adopted the mantra of referring to growth in GNP as the fundamental goal for their society and for their efforts in governing it, and even stranger that the voters do not question this mantra. Or perhaps it is not so surprising, considering that the political trick employed is to threaten people that without growth they shall lose their jobs and be unable to put food on the table. From this, it may be

Energy, Resources and Welfare. http://dx.doi.org/10.1016/B978-0-12-803218-3.00003-3

worthwhile to step a little deeper into the background behind the claimed connection between economic growth and welfare constituents such as being able to live a fulfilling life without lacking basic necessities. The further question of how economic growth might be made compatible with sustainability is occasionally discussed (e.g., Arrow et al., 2012), but really boils down to the obvious statement that growth exploiting any kind of depletable resources is necessarily unsustainable, as is growth using renewable resources above the natural upper limit of regeneration (think of wind or solar energy or biomass). Only growth in nonmaterial categories of economic activity can in principle grow unlimited, independent of whether they contribute to welfare or not.

It should immediately be noted that the politicians preaching about growth or job loss are deliberately acting contrary to the UN Human Right Charter, Article 23 (see Section 2.3.1), according to which it is their mission to ensure that citizens have jobs and get sufficient remuneration to allow a dignified and fulfilling life. This means that growth in activity is not desirable if it cannot be reconciled with arrangements where all citizens can enjoy work and leisure in a way that support the conditions for welfare. Economic growth is an acceptable goal to pursue only if it is consistent with creating conditions that maintain or improve welfare for everyone, not just for a few. The conditions for such welfare certainly include ensuring that wealth distribution is fair and disparity modest. The neo-liberal paradigm of maximizing the disparities between layers of society is not compatible with basic human rights.

As discussed in Section 2.2.1, the hope of neo-liberal advocates is that the poorer strata of society (or of nations) do not notice the increasing disparity if their own situation does not become worse, and if it does (e.g., during the recurrent financial collapses and their toll of job losses), that people should blame themselves rather than the economic system and the political leaders allowing the increases in economic disparity. Creating the notion of (nearly uninterrupted) growth is an important tool in maintaining this illusion. When the growth indicator includes both increases in the production of goods and services, growth in empty financial transactions and pollution, resource depletion and environmental degradation, then the growth will rarely halt and the population can be presented with daily news telling how well the nation is doing. The absent policy measures to ensure a fair distribution of the profits associated with the upward trend of the growth indicator can be played down, or at least so it would seem, due to the advertising campaigns and staged political discussions of petty matters in all media addressing the public, taking all the time and space available so that 'sorry, we do not have time to discuss distribution issues, because our amateur singing/dancing shows take all our time' or 'sorry, we get so many letters from the readers each day that we cannot possibly publish your comment on wealth distribution'. The following sections will discuss how things might change if the GNP growth indicator were replaced with a more relevant indicator of activity.

Growth may be said to be necessary for the poorer inhabitants of the world. More precisely, development is needed and in some cases, it may involve growth in resource use and thus interfere with the ongoing resource usage of richer countries, in case limiting factors are already in sight. This raises a number of questions, including what the sustainable human population of the world is. Of course it is not a constant but depends on the technology available and the efficiency by which it is used. However, the parts of the world that already use many resources inefficiently would likely need negative growth in resource exploitation (but possibly combined with positive growth in technological skills for efficient use of resources) in order to make room for nations lacking basic ingredients of well-being. How this may be achieved is, unfortunately, not a part of the current economic paradigm. United Nation's organizations such as the World Bank (2006) or UN (2006–2014) study group set goals for development of the poorer parts of the global economy, but rarely address the adjustments necessary in the affluent parts of the world, and seem to shy away from any discussion of population stabilization. It may happen by itself if the poor get rich, but history teaches us that this takes time and that the population levels resulting from such transitions are about an order of magnitude higher than at the starting point, which may not be acceptable in the current situation. Not long ago, a substantial interest in the relation between population size and economic growth was voiced (e.g., Nordhaus and Tobin, 1972).

In closing this section, it seems fair to quote John Galbraith in one of his more sarcastic appraisals of his own profession, 'In economics, it is professionally better to be associated with highly respectable error than uncertainly established truth' (Galbraith, 1976).

3.1.2 Measure of Desirable Activities

The convenience provided by the simplicity of the GNP is perhaps the reason that more meaningful indicators have not gained acceptance. Here, I shall explore an alternative that is nearly as simple to collect and work with, but which of course also can be criticised for the simplifications made. It is certainly not the first time that the suggestion of alternatives to GNP has been made. Following the oil crisis in late 1973, economists proposed valuing the exhaustion of fossil reserves, just as produced equipment is depreciated (Weltzman, 1976). Comments and an extension to a more general account of natural resources, depletable or not, were later added (e.g., Solow, 1986; Hartwick, 1990), typically using highly simplified mathematical toy models to derive obvious relations between assumptions and outcomes. In recent years, further concern has evolved around the pollution and climate change caused by the activities contributing to the GNP, and the possible aversion measures taken and also being counted by the GNP.

International organizations have stressed that both resource depletion and environmental damage should somehow be subtracted from the GNP (or, seen as more convenient by the economists mentioned earlier, from the net national product) in what is sometimes denoted a 'green GNP'. Notably, a group of international organizations have developed a framework for environmental accounting, making it closely resemble the traditional economic accounting by representing land and other durable property through a rental value, as suggested in Section 2.1.1, but at the same time not questioning private ownership of such assets (UN, EU, FAO, IMF, OECD, WB, 2014) and tacitly assigning zero economic value to natural assets not being part of any economic activities. These limitations are the result of wishing to make the environmental–economic accounting framework as similar as possible to the framework for conventional national accounting developed earlier by essentially the same organizations (EC, IMF, OECD, UN, WB, 1993). However, the international organizations closest to actual politics have generally taken the stand that incorporating issues of environmental sustainability in the GNP would be a lot of work, because many of the necessary data are not collected (Stieglitz et al., 2009*), and therefore they instead suggest, at least for the time being, that indicators for environmental and social impacts be defined and assessed in nonmonetary units and subsequently submitted to political decision makers along with the traditional GNP (e.g., EC, 2013; OECD, 2014; a short summary can be found in Gravgård, 2013).[†] This would make sure that the GNP, at least for a while, can continue to be heralded as a goal in political campaigns, while the new indicators perpetually play a secondary role not associated with compulsory political action, and are mainly there just to show the voters that the issues involved have not been forgotten. The difference between activities and welfare is in any case clear from the absence of activities not paid for in the GNP (work in own household and garden, free work for charity and sports organizations, and so on). Not only does this exclude using GNP as a measure of welfare, it also prevents identification of developments changing in time, such as the transfer of care for children and the

---

*Commendably, the report starts by reminding that the economic activity measure is not a measure of wellbeing and that failing to recognize this can lead to wrong political decisions.

[†]The US Government does not seem to have formally addressed these issues, except for Environmental Protection Agency advice to private companies wishing to use green accounting in marketing. Private initiatives include World Resources Institute (2003).

elderly from the unpaid family sphere to enterprises within the economic web. Quite generally, the activities that people consider as contributing most to welfare are rarely the activities of highest cost, as measured by GNP or any other monetary activity measure. It is therefore in any case necessary first to repair the monetary activity measure to reflect only activities desirable for society (by excluding activities such as polluting or destroying finite resources or moving funds around in circles), and then to discuss the translation of each activity into a measure of the associated welfare, if any.

In contrast to many of the earlier initiatives, which propose to treat environmental changes (in natural resources and induced from pollution) by an overall subtraction from the GNP, the consideration here will be the alternative of proceeding by an item-by-item evaluation, looking at the positive and negative impacts of each activity, not only from a life-cycle perspective but for any extent of time where impacts may happen. This idea grew out of discussions during the 1970s with Swedish aeronautical engineer Olle Ljungström, and may be called a measure of desirable activities (MDA) (Chapter 14 in Sørensen, 2014), because it divides the components of the GNP activities into three general categories: those predominantly desirable to society and its members, those more or less neutral (no benefit but likely also no harm) and finally those that are damaging to environment or society or any of its members. Each component of the GNP would be multiplied by a weight factor that is positive, zero or negative in the three cases, and finally summed up in case one wants a single number to replace the GNP, rather than a detailed account. The weighted sum would clearly be much closer to an indicator of wealth than the GNP summation with unit weight for all activities, even though the precise determination of the weight factors is of course uncertain and sometimes depending on arguments and valuations on which citizens may disagree. In its simplest form, the MDA would use a weight factor of '1' for desirable activities, '0' for neutral activities and '−1' for damaging activities.

In practice, the weighing would be done in a slightly more complex way, because the data used for calculating a country's GNP are available in terms of input–output matrices, telling how each category of import is distributed to manufacturers and service providers, and then how each step of industrial processing, assembly, conversion and despatch between enterprises is implying payments between companies in a matrix form. Finally, how each good or service is sold to private and public consumers, possibly enlarging the stocks of certain assets, or is exported. The negative social and environmental impacts that affect the monetary value of each transaction or activity chunk may be described as a monetary value to be subtracted from each particular activity represented by a point in the input–output matrix,

I_{jk} = environmental flow impacts (emissions to air, soil and water during all stages from materials extraction over production, distribution, usage and final disposal)

+ environmental stock impacts (mineral depletion, recycling recovery)

+ social impacts (health issues, work conditions, etc.)

$= c_{jk} \, i_{jk},$

where the index 'j' denotes production/service sectors and the index 'k' denotes output sectors of private or public consumption, stock change or export. The last equality indicates the possibility of writing the impact to subtract at each input–output step as a factor multiplying the GNP contribution by the activity step in question, c_{jk}. Some of the impacts relate more closely to the production or service-providing phase (sector 'j'), while others relate more closely to the use and disposal phase (in sector 'k'), allowing the impact factors to be written as

$$i_{jk} = e_j + f_k,$$

where e_j and f_k are weight factors applying to each sectors of input or output, so that the overall monetary evaluation of the national activities with consideration of social and environmental impacts becomes

$$MDA = \sum_{j,k} c_{jk}(1 - e_j - f_k).$$

The conventional GNP is obtained by making e_j and f_k equal zero. The translation of MDA into a measure of welfare (although only the components from activity flows, not from property or other stocks of wealth) would require an additional factor w_k for each end-use activity, to be determined, preferably, by direct expression of opinion by the members of society themselves,

$$\text{Welfare} = \sum_{k} w_k \sum_{j} [c_{jk}(1 - e_j - f_k)] + \text{components with no cost}$$

$$+ \text{stocks of wealth, weighed}.$$

The weight factor, w_k, differentiates otherwise positive activities according to how much they benefit society and its individual members, rather than to how much they cost, and expresses an attitude toward the severity of those impacts that have been found to be negative. High welfare contributions may come from activities with small monetary value, as well as from activities without monetary assignment, in which case they would have to be added to the restricted data from the national accounts, as indicated in the formula given earlier. The stocks of wealth may also have to be weighed in accordance with personal preferences, averaged over the population of the society for which the MDA is evaluated.

One may determine all the weights or at least the welfare weights by democratic vote, where each member of society can examine the list of activities (on the Internet, similar activities possibly grouped together in order to limit the overwhelming number of categories used in many current national accounts) and place his or her weight factors e_j, f_k and w_k (using a personal identity code as already required for all communication with government entities in many countries). The average of the weight factors for a particular activity from all those voting would be the one to use in the MDA. Clearly, a particular citizen does not need to put a weight factor on each component of activity, but can restrict the entries to activities for which the voter has views and feel confident putting in a number. Also, this would allow citizens to change their voting at any time if new information or insights induces them to change their ranking of any particular type of activity. The number of votes cast for each activity category, plus average and the standard deviation of voted weight factors for the category, can be publicized as assistance to further voting actions. As with any voting over the Internet,[‡] strict security measures would have to be in place to avoid misuse.

Naturally, an alternative approach would be to have weight factors determined by independent scientists with deep insights in the scientific fields affecting each weight factor assignment. Such an approach, say, with 'experts' placed in government committees, or with departmental civil servants instead of independent scientists, would be similar to the way such enquiries are being conducted at present in many countries. This opens the usual discussions of whether 'official' experts are free of influence from lobby groups or from the political preferences of the sitting government, just like the use of general referenda or direct democracy via the Internet opens questions of whether citizens have sufficient knowledge to choose the weight factors in question, and whether they can perhaps be manipulated by interest groups to vote in ways not reflecting their own basic interests, by making use of subversive hiding of underlying mechanisms determining the real implications of casting votes in a particular way. More on such kinds of direct democracy follows in Section 3.2.4, as well in the scenario in Chapter 5.

As an example, the monetary quantities needed for determinating the MDA have been calculated for the Danish input–output tables of the 2011 national account, using a quickly created set of subjective weight factors, e_j and f_k, not seriously subjected to scrutiny, in order to illustrate the procedure. Table 3.1 shows the effect of applying the reductions in 'green activity

---

[‡]The 'Internet' may not refer to the current implementation that is increasingly being hijacked by commercial advertisers and hackers, but to any future, hopefully improved system.

TABLE 3.1 Example of Applying Corrections Due to Resource Depletion and Environmental and Social Impacts to the National Activity Accounts of Denmark for 2011. The fraction shown in the column 'Impact factor' is for each sector of production or service applied to the corresponding activity figure in monetary terms and deducted to get the values given in the two last columns, representing output consumption and stock change split according to whether the input is import or a local enterprise. 1 DKK is about €0.13 or US $ 0.15 (mid-2015).

Monetary evaluation of national activities		Private and public consumption and stock changes. Environmental and social impacts in production and service phases deducted	
Denmark 2011, production/service sector	Impact factor	Import origin (M DKK)	National source (M DKK)
Agriculture and horticulture	0.5	743	1,680
Forestry	0.3	42	1,191
Fishing	0.3	92	95
Extraction of oil and gas	0.9	−80	−147
Extraction of gravel and stone	0.5	74	79
Mining support service activities	0.5	0	111
Production of meat and meat products	0.7	486	1,935
Processing and preserving of fish	0.1	640	410
Manufacture of dairy products	0.1	1,610	5,995
Manufacture of grain mill and bakery products	0.1	1,480	5,280
Other manufacture of food products	0.1	6,958	4,728
Manufacture of beverages	0.1	2,182	3,498
Manufacture of tobacco products	0.1	413	−162
Manufacture of textiles	0.3	643	466
Manufacture of wearing apparel	0.4	3,610	319
Manufacture of leather and footwear	0.5	753	55
Manufacture of wood and wood products	0.3	743	446
Manufacture of paper and paper products	0.5	444	47
Printing, etc.	0.8	19	10
Oil refinery, etc.	0.8	539	2,194
Manufacture of basic chemicals	0.7	32	255

(Continued)

TABLE 3.1 Example of Applying Corrections Due to Resource Depletion and Environmental and Social Impacts to the National Activity Accounts of Denmark for 2011. The fraction shown in the column 'Impact factor' is for each sector of production or service applied to the corresponding activity figure in monetary terms and deducted to get the values given in the two last columns, representing output consumption and stock change split according to whether the input is import or a local enterprise. 1 DKK is about €0.13 or US $ 0.15 (mid-2015). (cont.)

Monetary evaluation of national activities		Private and public consumption and stock changes. Environmental and social impacts in production and service phases deducted	
Denmark 2011, production/service sector	Impact factor	Import origin (M DKK)	National source (M DKK)
Manufacture of paints and soap, etc.	0.7	215	324
Pharmaceuticals	0.5	3,117	4,791
Manufacture of rubber and plastic products	0.6	169	385
Manufacture of glass and ceramic products	0.3	193	46
Manufacture of concrete and bricks	0.4	82	314
Manufacture of basic metals	0.5	177	157
Manufacture of fabricated metal products	0.5	1,280	713
Manufacture of computers and communication equipment, etc.	0.6	3,217	631
Manufacture of other electronic products	0.6	2,840	2,056
Manufacture of electric motors, etc.	0.6	162	342
Manufacture of wires and cables	0.6	83	33
Manufacture of household appliances, lamps, etc.	0.5	1,262	480
Manufacture of engines, windmills and pumps	0.5	527	3,333
Manufacture of other machinery	0.5	5,520	5,195
Manufacture of motor vehicles and related parts	0.5	8,041	296
Manufacture of ships and other transport equipment	0.5	139	393
Manufacture of furniture	0.5	2,170	1,339
Manufacture of medical instruments, etc.	0.5	581	579
Manufacture of toys and other manufacturing	0.5	1,150	1,236

(Continued)

TABLE 3.1 Example of Applying Corrections Due to Resource Depletion and Environmental and Social Impacts to the National Activity Accounts of Denmark for 2011. The fraction shown in the column 'Impact factor' is for each sector of production or service applied to the corresponding activity figure in monetary terms and deducted to get the values given in the two last columns, representing output consumption and stock change split according to whether the input is import or a local enterprise. 1 DKK is about €0.13 or US $ 0.15 (mid-2015). (*cont.*)

Monetary evaluation of national activities		Private and public consumption and stock changes. Environmental and social impacts in production and service phases deducted	
Denmark 2011, production/service sector	Impact factor	Import origin (M DKK)	National source (M DKK)
Repair and installation of machinery and equipment	0.4	945	436
Production and distribution of electricity	0.3	445	7,855
Manufacture and distribution of gas	0.4	65	2,010
Steam and hot water supply	0.3	0	6,077
Water collection, purification and supply	0.3	0	2,127
Sewerage	0.3	0	4,683
Waste management and materials recovery	0.3	15	3,963
Construction of new buildings	0.6	76	19,614
Civil engineering	0.1	861	32,421
Professional repair and maintenance of buildings	0.6	56	17,125
Own account repair and maintenance of buildings	0.6	9	2,934
Sale of motor vehicles	0.2	0	13,020
Repair and maintenance of motor vehicles, etc.	0.7	0	3,621
Wholesale	0.3	4	34,713
Retail sale	0.4	8	48,757
Passenger rail transport, interurban	0.5	31	2,208
Transport by suburban trains, buses and taxi operation, etc.	0.7	30	2,945
Freight transport by road and via pipeline	0.8	13	109
Water transport	0.2	18	1,504

(*Continued*)

TABLE 3.1 Example of Applying Corrections Due to Resource Depletion and Environmental and Social Impacts to the National Activity Accounts of Denmark for 2011. The fraction shown in the column 'Impact factor' is for each sector of production or service applied to the corresponding activity figure in monetary terms and deducted to get the values given in the two last columns, representing output consumption and stock change split according to whether the input is import or a local enterprise. 1 DKK is about €0.13 or US $ 0.15 (mid-2015). (*cont.*)

Monetary evaluation of national activities		Private and public consumption and stock changes. Environmental and social impacts in production and service phases deducted	
Denmark 2011, production/service sector	Impact factor	Import origin (M DKK)	National source (M DKK)
Air transport	0.9	68	41
Support activities for transportation	0.5	41	2,749
Postal and courier activities	0.8	0	168
Hotels and similar accommodation	0.4	0	3,918
Restaurants	0.4	0	18,548
Publishing	0.4	308	3,778
Publishing of computer games and other software	0.2	990	1,954
Film, television and sound recording activities	0.4	0	2,870
Radio and television broadcasting	0.2	7	3,652
Telecommunications	0.2	1,484	13,362
Information technology service activities	0.2	2,603	7,576
Information service activities	0.2	177	727
Monetary intermediation	0.8	1	5,650
Mortgage credit institutes, etc.	0.7	0	4,503
Insurance and pension funding	0.6	5	7,551
Other financial activities	0.8	0	388
Buying and selling of real estate	0.8	0	1,284
Renting of nonresidential buildings	0.2	0	419
Renting of residential buildings	0.2	0	51,325
Owner-occupied dwellings	0.2	0	84,073

(*Continued*)

TABLE 3.1   Example of Applying Corrections Due to Resource Depletion and Environmental and Social Impacts to the National Activity Accounts of Denmark for 2011. The fraction shown in the column 'Impact factor' is for each sector of production or service applied to the corresponding activity figure in monetary terms and deducted to get the values given in the two last columns, representing output consumption and stock change split according to whether the input is import or a local enterprise. 1 DKK is about €0.13 or US $ 0.15 (mid-2015). (*cont.*)

Monetary evaluation of national activities		Private and public consumption and stock changes. Environmental and social impacts in production and service phases deducted	
Denmark 2011, production/service sector	Impact factor	Import origin (M DKK)	National source (M DKK)
Legal activities	0.8	62	693
Accounting and bookkeeping activities	0.6	1	115
Business consultancy activities	0.3	0	766
Architectural and engineering activities	0.3	33	2,820
Scientific research and development (market)	0.2	3,458	5,614
Scientific research and development (non-market)	0.2	0	2 856
Advertising and market research	0.8	0	22
Other technical business services	0.8	0	101
Veterinary activities	0.3	0	777
Rental and leasing activities	0.6	4	699
Employment activities	0.5	5	5 779
Travel agent activities	0.6	0	5,546
Security and investigation activities	0.8	0	14
Services to buildings, cleaning and landscape activities	0.8	0	520
Other business service activities	0.8	0	120
Public administration	0.5	0	38,405
Defence, public order, security and justice activities	0.8	0	9,324
Rescue service etc. (market)	0.6	4	595
Primary education	0.2	0	44,881

(*Continued*)

TABLE 3.1 Example of Applying Corrections Due to Resource Depletion and Environmental and Social Impacts to the National Activity Accounts of Denmark for 2011. The fraction shown in the column 'Impact factor' is for each sector of production or service applied to the corresponding activity figure in monetary terms and deducted to get the values given in the two last columns, representing output consumption and stock change split according to whether the input is import or a local enterprise. 1 DKK is about €0.13 or US $ 0.15 (mid-2015). (*cont.*)

Monetary evaluation of national activities		Private and public consumption and stock changes. Environmental and social impacts in production and service phases deducted	
Denmark 2011, production/service sector	Impact factor	Import origin (M DKK)	National source (M DKK)
Secondary education	0.2	0	22,891
Higher education	0.2	0	29,283
Adult and other education (nonmarket)	0.2	0	4,297
Adult and other education (market)	0.2	271	1,567
Hospital activities	0.5	48	42,401
Medical and dental practice activities	0.5	57	16,514
Residential care activities	0.5	0	15,192
Social work activities without accommodation	0.3	0	67,718
Theatres, concerts and arts activities	0.3	139	5,604
Libraries, museums and other cultural activities (market)	0.2	0	593
Libraries, museums and other cultural activities (nonmarket)	0.2	0	8,485
Gambling and betting activities	0.8	0	826
Sports activities (market)	0.5	13	1,143
Sports activities (nonmarket)	0.5	0	2,593
Amusement and recreation activities	0.7	0	999
Activities of membership organizations	0.8	0	4,124
Repair of personal goods	0.4	1	1,972
Other personal service activities	0.5	62	4,161
Activities of households as employers of domestic personnel	0.5	0	2,340

Source: Basic input–output data are from Statistics Denmark (2015a).

measures' derived from resource depletion and environmental and social impacts, but only to the production and service providing sectors, not to the consumption sectors. Tables 3.2 and 3.3 include both impacts on the 'green activity measure' arising within the production/service sectors and also those from the consumption sectors and stock changes. In Table 3.2 the effect on private consumption sectors is depicted and in Table 3.3 the effect on public consumption sectors. The Statistics Denmark (2015a) categories used in the input–output matrix divide production and service into 235 categories and consumption into 295 categories. This is far too aggregated to allow a precise determination of the impact numbers to be subtracted. It is not known if the agricultural activities use ecological (organic) cultivation methods or traditional chemical cultivation methods with pesticides, etc. Similarly, the tables do not indicate which producers or service providers use renewable energy rather than fossil or nuclear fuels, and the tables cannot identify producers that use mineral resources in danger of depletion. Some of these data may be found in other statistical bases. For instance, the known penetration of ecological farming in Denmark (Danish Ministry of Food, Agriculture and Fisheries, 2010; Sørensen, 2012) and the share of renewable energy sources (Statistics Denmark, 2015b) are available and have been considered in the choice of the weight factors shown in Table 3.1. However, in many cases, such as for soil or waterway pollution and the use of rare minerals, the available data is of limited use (for example, statistics on materials use broad categories of physical amounts, while toxicity and emissions have rarely been quantified in more than a few specific case studies). In order to proceed from statistics of physical quantities, the monetizing of health and social effects from emissions have to use available life-cycle studies to quantify impacts (EC, 1995; Sørensen, 2011). Most impacts do not respect borders, in particular, the impacts from greenhouse gas emissions are felt globally, but with differences depending on local climatic and social conditions.

To arrive at the GNP (all weight factors equal to one) or the MDA (including subtractions due to impacts from resource depletion, environmental degradation and negative social impacts), one has to sum all the consumption and stock change sector activities, add the exports and subtract the imports. Using the Danish 2011 national accounts as an example, the value of the exports are 12% above that of the imports, but the MDA difference resulting from the impact corrections is much larger, because for imports, corrections are made both for production and use, but for exports only for production. This is not an error, because the impacts from the use of goods and services exported by Denmark should be incorporated in the green accounts of the countries that import them, and not in the Danish account. A dominating contribution to the Danish CO_2 emissions is from air and particularly ship emissions outside Denmark, due to the country's large presence in international sea hauling of goods. These emissions are

TABLE 3.2 Example of Applying Corrections Due to Resource Depletion and Environmental and Social Impacts to the National Activity Accounts of Denmark for 2011. The fraction shown in the column 'Consumption impact factor' is for each sector of consumption or stock change. Together with the impact factor of Table 3.1, it is multiplied by and subtracted from the corresponding activity figure in monetary terms, to give the result shown in the last column, here only for private sector outputs.

Monetary evaluation of national activities Denmark 2011, private sector	Consumption impact factor	Private consumption (both production and use impact estimates subtracted) (M DKK)
Bread and cereals	0.2	8,600
Meat	0.4	2,708
Fish	0.3	1,617
Eggs	0.3	471
Milk, cream, yoghurt, etc.	0.3	3,942
Cheese	0.4	2,056
Oils and fats	0.5	1,036
Fruit and vegetables except potatoes	0.3	5,661
Potatoes, etc.	0.2	956
Sugar	0.4	184
Ice cream, chocolate and confectionery	0.3	6,603
Food products	0.3	2,127
Coffee, tea and cocoa	0.5	1,014
Mineral waters, soft drinks, fruit and vegetable juices	0.5	2,534
Spirits and wine	0.5	3,397
Beer	0.5	1,591
Tobacco, etc.	1.0	−2,864
Articles of clothing	0.3	11,570
Cleaning, repair and hire of clothing	0.6	7
Footwear	0.2	3,314
Actual rentals for housing	0.2	38,118
Imputed rentals for housing	0.2	63,322
Maintenance and repair of the dwelling	0.4	925
Refuse collection, other services	0.5	3,845
Water supply and sewerage services	0.5	1,135

(Continued)

TABLE 3.2   Example of Applying Corrections Due to Resource Depletion and Environmental and Social Impacts to the National Activity Accounts of Denmark for 2011. The fraction shown in the column 'Consumption impact factor' is for each sector of consumption or stock change. Together with the impact factor of Table 3.1, it is multiplied by and subtracted from the corresponding activity figure in monetary terms, to give the result shown in the last column, here only for private sector outputs. (*cont.*)

Monetary evaluation of national activities		Private consumption (both production and use impact estimates subtracted) (M DKK)
Denmark 2011, private sector	Consumption impact factor	
Electricity	0.3	12,909
Gas	0.7	758
Liquid fuels	0.9	−1,498
District heating, etc.	0.6	4,368
Furniture and furnishings, carpets and other floor coverings	0.5	2,608
Household textiles	0.5	656
All major household appliances and small electric household appliances	0.5	943
Repair of major household appliances	0.5	52
Glassware, tableware and household utensils	0.3	1,641
Tools and equipment for house and garden	0.5	712
Nondurable household goods	0.6	350
Domestic services and household services	0.4	153
Pharmaceutical products and other medical products	0.3	2,670
Therapeutic appliances and equipment	0.3	1,215
Out-patient services	0.4	1,081
Hospital services	0.4	375
Purchase of vehicles	0.3	12,977
Spare parts, maintenance and repair of personal transport equipment	0.6	−2,603
Fuels and lubricants for personal transport equipment	1.0	−7,799
Other services in respect of personal transport equipment	0.9	−2,840

(*Continued*)

TABLE 3.2 Example of Applying Corrections Due to Resource Depletion and Environmental and Social Impacts to the National Activity Accounts of Denmark for 2011. The fraction shown in the column 'Consumption impact factor' is for each sector of consumption or stock change. Together with the impact factor of Table 3.1, it is multiplied by and subtracted from the corresponding activity figure in monetary terms, to give the result shown in the last column, here only for private sector outputs. (*cont.*)

Monetary evaluation of national activities Denmark 2011, private sector	Consumption impact factor	Private consumption (both production and use impact estimates subtracted) (M DKK)
Transport services	0.9	−9,484
Postal services	0.7	−249
Telephone and data communication equipment	0.4	−25
Telephone and data communication services	0.5	4,937
Radio and television sets, etc.	0.4	1,658
Photographic equipment, etc.	0.5	154
Data processing equipment	0.4	2,036
Recording media for pictures and sound	0.4	778
Repair of a/v and data processing equipment	0.7	−87
Other major durables for recreation and culture	0.4	1,060
Other recreational items and equipment, gardens and pets	0.7	−316
Recreational and cultural services	0.1	16,884
Books, newspapers, periodicals and miscellaneous printed matter	0.6	474
Stationery and drawing materials, etc.	0.4	280
Package holidays	0.8	−5,330
Education	0.1	4,360
Restaurants and catering	0.8	−5,517
Accommodation services	0.5	1,236
Hairdressing salons and personal grooming establishments	0.7	−824
Appliances, articles and products for personal care	0.3	3,153

(*Continued*)

TABLE 3.2   Example of Applying Corrections Due to Resource Depletion and Environmental and Social Impacts to the National Activity Accounts of Denmark for 2011. The fraction shown in the column 'Consumption impact factor' is for each sector of consumption or stock change. Together with the impact factor of Table 3.1, it is multiplied by and subtracted from the corresponding activity figure in monetary terms, to give the result shown in the last column, here only for private sector outputs. (*cont.*)

Monetary evaluation of national activities		Private consumption (both production and use impact estimates subtracted) (M DKK)
Denmark 2011, private sector	Consumption impact factor	
Jewellery, clocks and watches	0.2	860
Other personal effects	0.3	1,189
Retirement homes, day-care centres, etc.	0.3	1,526
Kindergartens, nurseries, etc.	0.3	4,384
Insurance	0.0	8,445
Financial services	0.4	−6,704
Other services	0.3	494
Consumption by nonresidents on the economic territory	0.7	7,542
Consumption by residents in the ROW	0.7	−7,464
Private hospital services	0.4	538
Private recreational and cultural services	0.2	1,546
Private education	0.1	7,247
Private retirement homes, day-care centres, etc.	0.3	717
Private kindergartens, nurseries, etc.	0.3	116
Other private services	0.2	0

Source: Basic input–output data are from Statistics Denmark (2015a).

not included in the traditional green accounting, as they primarily relate to exports of transport services.

The reduction from GNP to MDA presented here cannot be compared with the preliminary attempts to establish a green GNP by Statistics Denmark (Gravgård, 2013), because Statistics Denmark only consider a limited devaluing of natural resources similar to that of production machinery, that is with use of a similar positive discount rate, prone to the criticism discussed in Section 2.1. Global climatic impacts are not considered at all. Pollution of soil and waterways is not available in monetary

TABLE 3.3 Example of Applying Corrections Due to Resource Depletion and Environmental and Social Impacts to the National Activity Accounts of Denmark for 2011. The fraction shown in the column 'Consumption impact factor' is for each sector of consumption or stock change. Together with the impact factor of Table 3.1, it is multiplied by and subtracted from the corresponding activity figure in monetary terms, to give the result shown in the last column, here only for public sector outputs. At the end, imports and exports are summed up, although no consumption impact subtraction can be done for the export values without doing similar studies for the countries receiving the exports. In any case, such impacts should, for consistency, be accounted for by the receiving countries, not by the exporting country.

Monetary evaluation of national activities		Public consumption (both production and
Denmark 2011, public sector	Consumption impact factor	use impact estimates subtracted) (M DKK)
Food products	0.3	93
Footwear	0.2	8
Maintenance and repair of the dwelling	0.4	38
Furniture and furnishings, carpets and other floor coverings	0.5	26
Household textiles	0.5	43
Domestic services and household services	0.4	322
Pharmaceutical products and other medical products	0.3	2,573
Therapeutic appliances and equipment	0.3	594
Out-patient services	0.4	1,534
Purchase of vehicles	0.3	183
Transport services	0.4	−39
Data processing equipment	0.1	13
Appliances, articles and products for personal care	0.3	3
Insurance	0	66
Therapeutic appliances and equipment	0.3	308
Out-patient services	0.4	846
Hospital services	0.4	4,082
Recreational and cultural services	0.1	6,894
Education	0.1	60,346
Retirement homes, day-care centres, etc.	0.3	11,715

(Continued)

TABLE 3.3 Example of Applying Corrections Due to Resource Depletion and Environmental and Social Impacts to the National Activity Accounts of Denmark for 2011. The fraction shown in the column 'Consumption impact factor' is for each sector of consumption or stock change. Together with the impact factor of Table 3.1, it is multiplied by and subtracted from the corresponding activity figure in monetary terms, to give the result shown in the last column, here only for public sector outputs. At the end, imports and exports are summed up, although no consumption impact subtraction can be done for the export values without doing similar studies for the countries receiving the exports. In any case, such impacts should, for consistency, be accounted for by the receiving countries, not by the exporting country. (*cont.*)

Monetary evaluation of national activities		Public consumption (both production and use impact estimates subtracted) (M DKK)
Denmark 2011, public sector	Consumption impact factor	
Kindergartens, nurseries, etc.	0.3	11,097
General services	0.2	6,692
Defence	1	−19,745
Order, safety	0.2	100
Economic affairs	0.5	2,540
Environmental protection	0.1	2,514
Housing, community amenities	0.4	212
Health	0.4	544
Recreation, culture and religion	0.2	2,069
Education	0.1	1,123
Social protection	0.3	2,027
Stock change		
Dwellings	0.4	8,056
Buildings other than dwellings	0.4	2,765
Other structures and land improvements	0.5	13,975
Transport equipment	0.6	4,491
Computer hardware	0.2	5,836
Telecommunication equipment	0.2	510
Other machinery and equipment and weapon systems	0.6	−2,752
Cultivated biological resources	0	−73
Research and development	0.1	27,676

(Continued)

TABLE 3.3 Example of Applying Corrections Due to Resource Depletion and Environmental and Social Impacts to the National Activity Accounts of Denmark for 2011. The fraction shown in the column 'Consumption impact factor' is for each sector of consumption or stock change. Together with the impact factor of Table 3.1, it is multiplied by and subtracted from the corresponding activity figure in monetary terms, to give the result shown in the last column, here only for public sector outputs. At the end, imports and exports are summed up, although no consumption impact subtraction can be done for the export values without doing similar studies for the countries receiving the exports. In any case, such impacts should, for consistency, be accounted for by the receiving countries, not by the exporting country. (*cont.*)

Monetary evaluation of national activities	Consumption impact factor	Public consumption (both production and use impact estimates subtracted) (M DKK)
Denmark 2011, public sector		
Mineral exploration and evaluation	0.4	−307
Computer software and databases	0.1	18,341
Entertainment, literary or artistic originals and other intellectual property products	0.1	2,892
Valuables	0	2,010
Inventories	0	7,797
Exports (without deductions for negative impacts during use)		548,106
Imports (with deductions for negative impacts in production and use)		371,262
'Green GNP' (misleading in respect to import/export ratio)		595,090
Consumption, stock change and export corrected for national impacts		966,352
Total stock change		91,216
Total private consumption		228,210
Total public consumption		98,819

Source: Basic input–output data are from Statistics Denmark (2015a).

terms, and neither are data for materials depletion (the data on materials maintained by Statistics Denmark only specify masses, with no distinction between gravel and gold). As a consequence, Statistics Denmark finds a modest reduction in GNP from 'green' corrections.

Fig. 3.1 shows the conventional GNP for Denmark in fixed prices from 1966 to 2011, along with other quantities derived from the input–output tables of Statistics Denmark (2015a). The GNP grows at a steady rate from 1966 to 1996, then nearly doubles its growth rate to 2006, after which

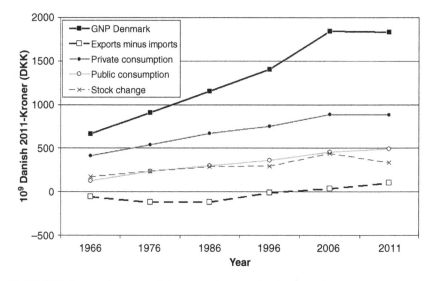

FIGURE 3.1 **Development of the GNP of Denmark, with some important constituents (in billion DKK at 2011 prices).** *Input–output data from Statistics Denmark (2015a) have been used.*

the 2007+ financial crises halts the growth and even exhibits a slight fall in GNP. The year 1996 is the turning point where a deficit in the foreign trade balance is replaced by a surplus, growing after the financial crisis. The consumption behavior shows that it is the private consumption that has stagnated after 2007, whereas public spending grows modestly throughout the 45-year period displayed. Some of the growth after 2006 (e.g., in exports) is accompanied by a decrease in stocks.

In Fig. 3.2, the corresponding behavior of the MDA is shown along with the deductions made by the crude illustrative model of MDA, on the same components as those shown in Fig. 3.1. First it is seen that the MDA grows much more rapidly than the GNP, and even grows after the financial crises. The reason for this is that the negative impacts subtracted to get from GNP to MDA have in a number of cases diminished during the period considered. Around 1960, the Danish agricultural production was based on extensive use of plant pesticides and preventive use of antibiotics on livestock animals. These abuses dropped by more than a factor of four from 1960 to 2010, and furthermore, ecological methods entered Danish agriculture in a major way from around the year 2000 (Sørensen, 2012). An even larger contributor to reducing negative impacts was the radical change of the energy industry, both by replacing coal by natural gas (with about half the greenhouse gas emissions) and by use of renewable energy such as the expansion of on- and off-shore wind power production. Fig. 3.2 shows that the decline in negative impacts from production is reflected in the MDA value

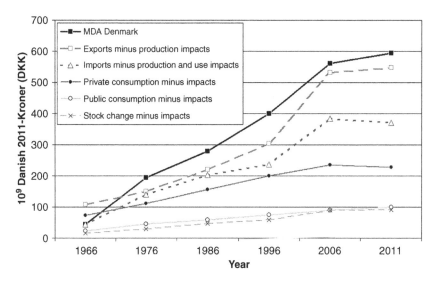

FIGURE 3.2 **Development of the MDA of Denmark with some important constitu-ents.** The MDA differs from the GNP by subtracting environmental and social impact costs in production and use, here based on a set of simplistic factors multiplying each activity transaction, as discussed in the text. These factors have been held constant except for the production impact factors of agriculture and electricity production, taking into account the declining use of pesticides and increasing ecological production, and the transition from oil and coal first to natural gas and then toward renewable energy power generation (currently covering about 40% of demand).

of exports. However, as noted, the impacts from usage are not deducted from exports because they must be evaluated in the country importing the goods (or services). Because usage impacts are deducted from imported goods and services; Fig. 3.2 no longer shows an approximate balancing between imports and exports. The private and public sector consumption shows the same trends as the corresponding GNP components, including the fall in private consumption. However, the value of stocks no longer declines after 2007, because the negative impacts associated with the changing stocks are smaller than at previous times. In summary, this small exercise shows that the MDA has the potential for a much better understanding of the real development of a national economy.

3.1.3 A New Look on Wealth and Inequality, Illustrated by a Case Study

While GDP or MDA are measures of monetary flows, wealth must be related to stocks of assets. Again, a case study for Denmark will be used, this time to exemplify an attempt to define a measure of wealth that takes into account the negative impacts, as discussed in the modification from GNP to

MDA. Regarding availability of data, the situation is poor compared with the flow analyses. Often, national statistical data on stocks are obtained from a time series of flow data, by accumulating additions and losses of assets over a period, but unfortunately, frequent changes in the definitions used for statistical variables make the summing up unreliable, in addition to the fact that some items have data only for a limited number of years.

The stocks of assets are of four kinds:

1. Natural assets (land, sea, forests, soils, underground and biota, where some of these are depletable (mineral deposits) or may be destroyed).
2. Produced assets (buildings, infrastructure, vehicles and communication means, machinery, electronics, etc., all with a finite lifetime).
3. Human assets (lives, skills, knowledge, cultural and social tools for creating coherence in a society, all of which are limited in duration by lifespan and have to be renewed).
4. Financial assets (tokens that may be exchanged for real assets, typically in the categories 2 or maybe 3).

In contrast to the World Bank concept of assets (capital) mentioned in Section 2.1.1, no unspecified 'rest category' is left as intangible. Those components for which a method of estimation has been identified are included. This comprises all the categories that the World Bank lists as "intangible".

The available Danish statistics enables an overview of some of the relevant assets, as depicted in Figs 3.3 and 3.4. Fig. 3.3 shows the sum of important produced assets (category 2), along with fossil assets (a depletable resource in category 1) and pension funds (category 4). Also indicated is the balance of financial instruments and two components of the passives: public and private debt. The private debt is one of the largest (per capita) in the world and some studies have expressed worry over the apparent stability of the Danish economy in the light of such debt levels (Schwartzkopff, 2013; Davies et al., 2009). The Danish National Bank has in response explained why the debt is not causing it concern (Andersen et al., 2012), first of all arguing that the pension funds build up by Danish citizens are also among the highest in the world, so that along with the welfare economy of the largely free medical, educational and to some extent social services, Danes do not have much motive not to consume as much as the money lenders will allow them, expecting the pension savings to ensure a satisfactory retirement life. The pension wealth is largely of the insurance type, and since the introduction of the welfare paradigm during the 1930s, it has been shaped by most salaried Danes paying 15% of their income to the pension operator. This has typically been 5% paid by the salaried person plus an additional 10% paid by the employer. This scheme was an important incentive when the system was introduced, because individuals

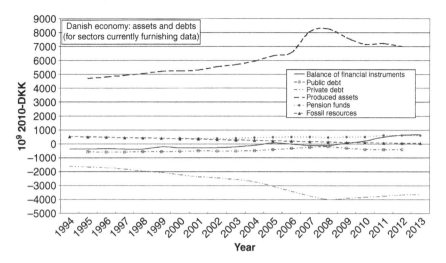

FIGURE 3.3 Selected stock inventory of the Danish economy in market prices adjusted to the 2010 price index. The balance of financial instruments covers all sector of the economy and both domestic and foreign assets and liabilities. The public debt comprises both domestic and international debts, while the private debt includes only domestic loans. The produced assets are detailed in Fig. 3.4. The pension funds are stock and equity holdings of private pension enterprises, as valued for each specific year, and the fossil resources are the sum of market values over the years remaining to depletion (the feasible production has already declined to a value that makes the estimate of extrapolated future extraction rather unimportant). *Data sources: Statistics Denmark (2015b,c,d); Skat (2015); Danish National Bank (2015).*

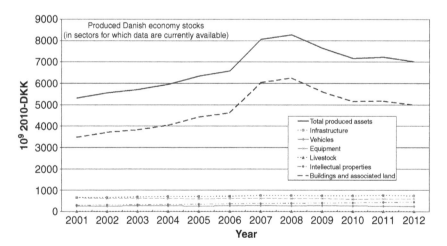

FIGURE 3.4 Stock inventory of some important produced assets in the Danish economy in market prices adjusted to the 2010 price index. The dominating values, those of building property, are taken from (currently biannual) taxation assessments (Skat, 2015), comprising all property with buildings (all types of dwellings, industrial, commercial and farm buildings). The value of the plot of land upon which the buildings are erected is included, but not the value of open land, forests, lakes or ocean areas. The remaining tangible assets are valued at purchase price minus depreciation along an assumed average lifetime, and a similar procedure is used for intellectual property (Statistics Denmark, 2015c). Infrastructure includes roads, bridges and railways, while equipment comprises any equipment used in commerce, industry and private households.

not wanting to part with 5% of their earnings also lost the 10% from the employer. Subsequently, the employer part has of course figured as part of the salary in union–employer negotiations.

The pension funds shown in Fig. 3.3 are the market values of the assets accumulated in the pension enterprises at present, that is stock and bond quotes, plus any other property of the pension enterprises, such as buildings. The very strict legislation regarding permitted pension investments in effect until 2 decades ago have been replaced by less strict rules. As a result, some pension enterprises lost considerable value in 2008–10 when stock values fell. Future pensions will depend on capable management of the pension funds, which may or may not materialize, and the current market value is therefore used here, rather than the predicted stock value with some projected interest earnings to the future start time of pension payments (on average 20–25 years) used by the National Bank in the 2012 paper quoted.

The Danish fossil resources, oil and natural gas from the North Sea, have been declining since their first exploitation around 1980. Because very few are left today and the extraction correspondingly declining, a credible stock curve is constructed based on an assumed depletion by early 2020s, by adding extracted amounts annually, backward to the beginning of the graph in 1994.

The largest assets depicted in Fig. 3.3 are those of produced goods including buildings. The assets valuated are detailed in Fig. 3.4. For the category 2 assets of vehicles, equipment, agricultural livestock, infrastructure such as roads, bridges and airports, the valuation is taken from Statistics Denmark (2015c), based on price at purchase or construction, adjusted by a standard rate of depreciation over an assumed lifetime. Also included, because they appear in the same context in the data sets used, is intellectual property (from the arts to technical patents, all belonging to category 3).

In the case of buildings, the similarly calculated data from Statistics Denmark are not used because this would hide market changes in value such as those seen after the burst of the bubble economy in 2007. It is a surprising fact that Danish houses are bought and sold at prices that do not decline systematically toward the end of life of the structure. Despite the compulsory condition and energy use reports that have to be produced when houses are traded, prices of houses at similar locations but with widely differing ranking are quite similar. In order to take such behavior into account, assessment values from the Danish tax authorities are used. They reflect actual sales prices in each neighborhood and comprise both the building (of any type from dwellings to agricultural, industrial and commercial buildings) and the plot of land on which it stands, thus stretching over categories 1 and 2. Other land resources without buildings are not included. The assessment for

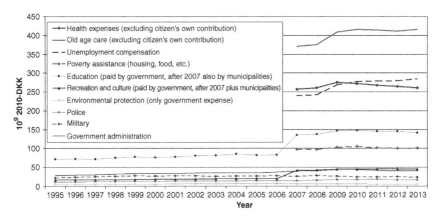

FIGURE 3.5 **Annual Danish expenditures for building or maintaining human and social capital (see text for details).** *Based on Statistics Denmark (2015c).*

taxation purposes takes place biannually (one year for dwellings, the next for other buildings) and aims to provide market prices. Therefore, the effect of the building price bubble's rise and fall is captured with only a slight delay.

Turning to human capital assets, Fig. 3.5 gives some background information on financial activities aimed to build up such assets, notably health and old age care, education and unemployment compensation, but also general expenses for administration, security, recreation and environmental protection. It is seen that the largest annual cost is for old age care, including help in homes (food delivery, cleaning, gardening, personal care) as well as in institutions (from especially equipped houses and apartments with access to mobile nurses to retirement homes and intensive care units). Next come health expenses for hospitals, specialized clinics and general practice doctors (dental work is strangely not supported except for certain age groups such as school children) and unemployment compensation. For poverty assistance the expenses include not just living expenses, but also aid in paying for dwelling rent or mortgage payments, such that for example a time-limited unemployment period does not force a family to move away from the dwelling they are accustomed to living in. The curve for educational expenses is discontinuous because before 2007, only government expenses are included, not the substantial municipal outlays, particularly for primary schools. The same is true for environmental protection, where the expenses carried by municipalities, for example for securing coastal and forest areas against storm damage or cleaning up after such damage, have not been included.

However, on a more general level of discussion, entities such as the capital health and education assets may not have any strong relation to the expenses spent on activities in these areas, and many of the social 're-pair' expenses, such as unemployment compensation, do not have any obvious function in building up human capital. First, education will be considered.

3.1.3.1 Education

In the most recent version of the World Bank study of wealth, education is included and represented by the expense that nations put into educa-tional activities (World Bank, 2011). As argued earlier, this is not an entirely fair representation of educational capital, so a different method of evalu-ation should be found. Educating individuals does not, of course, create a permanent stock of assets, but in the current, highly technology-oriented societies, it has to be renewed by further education, at least through the working life of each person, and then has to be passed on to the following generation by new education or other learning avenues.

An idea of the value of education may be obtained from the differences in earnings for people with different education. Again using Denmark as a case study, Fig. 3.6 shows the development in the distribution of the numbers of people attaining a given level of education, and Fig. 3.7 shows the income they on average is able to derive from their level of education. Because it takes longer to reach a higher level of education, it is some-times argued that the subsequent number of years with a higher income is smaller than for those with less education. This is only partially right, be-cause unskilled jobs (in present societies) are on average more physically demanding and lead to earlier attrition. As a result, people with higher education have a tendency to retire later, and therefore the lifelong earn-ings are not reduced as much as indicated by the length of study. It is inter-esting that Fig. 3.6 shows that the number of people with higher education has increased dramatically during the last decades, while Fig. 3.5 showed that the expenses devoted to education have stayed constant or slightly declined. One could marvel over the increase in educational productivity, but the real effect of the budget cuts relative to the number of students have also included dilution of the depth of curriculum, in addition to the obvious decline in individual student–teacher contact.

Fig. 3.6 indicates that the number of people with only primary school education is declining in Denmark, the number of people choosing voca-tional training is levelling off, and the number of people taking medium and long (academic) educations are on the rise. The income implications shown in Fig. 3.7 show that when both earned and subsidized income are counted, people with a vocational education earn 50% more than those with only the compulsory schooling, those with medium–high education have, since about the year 2000, earned around 80% more and people with an academic education earned on average 180% more, a picture that is only modestly altered by considering lifelong accumulated earnings. It

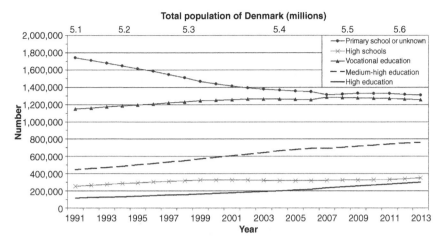

FIGURE 3.6   Highest level of formal education reached by the Danish population (total population at top). *Based on Statistics Denmark (2015e).*

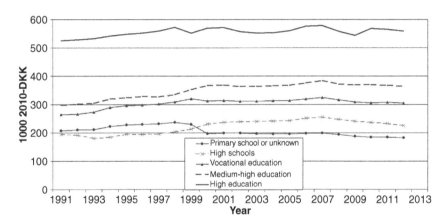

FIGURE 3.7   Average income per capita in Denmark (from salary, subsidy or own business) as a function of the highest educational level reached. Only persons in the age interval 15–65 are included. Subsidies such as unemployment benefits are included, and their size can be seen in Fig. 3.8. Pensions are not included. *Based on Statistics Denmark (2015f).*

should be noted that the statistical data used comprises all persons aged 15–65 years but counts income only from those with an earned income, either salaried and self-employed people or people living off interest/insurance or subsidy.

The implication of including income by public subsidy follows from Fig. 3.8, which distributes the subsidy payments on the same grouping of educational levels. Here it is seen that the group with only primary school education on average receives half of its income from subsidies and thus

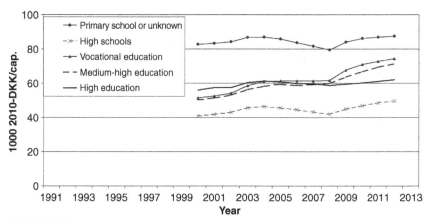

FIGURE 3.8 Level of public support (unemployment compensation and social contributions such as poverty assistance) averaged over all persons in each educational level group (ie not just over those receiving the support). *Based on Statistics Denmark (2015f).*

only half from their own work. For the higher education groups, the subsidy part is considerably smaller and for the academic level only 10–11% of the income has come from subsidies during the period 2000–12.

In order to calculate the value of the educational effort one might take the increase in earning for each person in society and multiply it by the average number of years through which the learning is expected to sustain a higher earning. In a future where lifelong education beyond the initial 'formal' education is more closely connected with increased income, more complex schemes of evaluation need to be used, but for the information presently available, using the difference between the earnings of each group from Fig. 3.7 and the subsidies of Fig. 3.8 to estimate the capital stock associated with further education will do. The earning difference should then be multiplied by 30–35, representing the years an average person of age just under 40 in each education group will continue to earn more than an unskilled person. Finally, a stock value for the assets of people with only primary school education should be added. This value is certainly not zero, as the primary school education in countries such as Denmark allows the recipients to perform many functions not possible by someone not having had primary school attendance. Setting the earning of the latter (rather hypothetical in a society like Denmark) at zero, the educational stock should just be calculated as indicated earlier, but relative to zero earning rather than relative to the earning of someone with only primary school education. This procedure along with an average knowledge lifetime at no depreciation of 35 years has been used in constructing the educational capital shown in Fig. 3.10. No depreciation is applied under the assumption that forgotten or outdated education is continuously renewed through professional experience and through adult education of

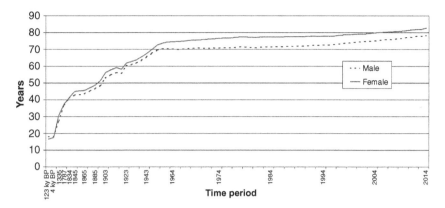

FIGURE 3.9 Variations in life expectancy at birth in Denmark (Sørensen, 2012; Statistics Denmark, 2015g). 'ky BP' is 1000 years before present.

suitable nature, whether taught or self-administered. The ability to update knowledge is fortunately becoming part of most current education, but unfortunately, many jobs are organized in ways that leave insufficient time for systematic and continued learning (ask your practising medical doctor what scientific journals he or she reads!).

3.1.3.2 Health

The consideration of citizens' health as an asset for society has been suggested by Arrow et al. (2012). Realizing that the expenditures of the health sector may be a poor indicator of value, Arrow et al. propose to measure the health as a capital asset by way of life expectancy, arguing that increased longevity is the outcome of medical activities. The 'zero-point' of this scale of value is not easy to define. What would the life expectancy be without health services? Hardly less than 20 years, like the life expectancy of early Stone Age humans (Sørensen, 2012). Life expectancy has risen as a result of changes in lifestyle, more adequate food selection, less dangerous work and many other things not part of the basic efforts of the health sector. In the case of Denmark, used earlier as an example, the average life expectancy at birth rose from around 44 years in around 1850 (for a while, poor living conditions characterizing early industrialization here halted the previous increase in life expectancy) to about 71 years by 1950, then showed little increase until the recent 2 decades with new life extensions from 75 to 81 years (Fig. 3.9). None of the changes correspond well with the changing amounts of money made available to the health sector, and in particular, the recent increase in life span has occurred in a period of declining health expenses (Fig. 3.5), suggesting lifestyle changes as a main cause (stopping smoking, eating less sugar and fat).

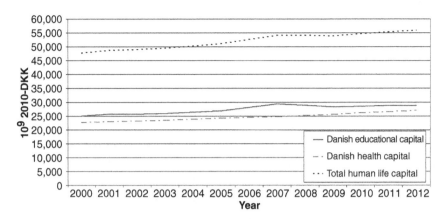

FIGURE 3.10   **Valuation of Danish human assets.** The total life capital is calculated as the sum of values describing the human capacity to service society through work, valued as earnings depending on education, and values of health and longevity extensions by medical and lifestyle advances, in turn permitting social coherence activities and cultural activities such as producing science and knowledge, art and entertainment. *Data are derived from the sources of Figs 3.8 and 3.9, as well as from EC (1995) and Sørensen (2011).*

It thus appears that if health sector expenses contribute to increased life span, they do so by only 10–20%. The fact that citizens have a given length of life can be seen as an asset to society, and its valuation could be divided into the value of the work each member of society can provide (and this is already quantified in the discussion of education and shown in Fig. 3.10, so it should not be double-counted as health capital), and a value attributed to other aspects of a citizen's life that may be termed contributions to social coherence. This again touches upon a discussion of the economic value of a human life that has been ongoing for a couple of decades.

In an effort to quantify externalities caused by pollutant emissions from power plants, the project ExternE of the European Commission (EC, 1995) used a value of a statistical life (VSL), lost due to adverse health effects outside the control of the victim, equal to €2.6 million in 1994 or about US$4.4 million in 2010. The basis for this highly uncertain value is a discussion of evidence for salary increases related to willingness to accept risky work, willingness to pay interviews and actual payments made to prolong life (antialcohol drugs, antismoking drugs, etc.). The most important determinant may well have been actual payments to victims of work-related deaths, such as the compensation paid by the French government to families of personnel dying while taking part in the Muroroa nuclear tests in the Pacific. In a discussion of these estimates, Sørensen (2011) first of all distinguishes between personal perceptions of the value of life, as revealed by life insurance payout and the indirect methods mentioned, and, on the other hand, the loss to society when a member accidentally dies. The societal definition should be used in economic discussions to avoid the criticism that losing a life has ethical dimensions that do not lend themselves easily to monetizing, and to some people should therefore not be discussed in prosaic terms.

The obvious loss to society when one of its members accidentally dies is losing the ability of that person to contribute work. This might be valued differently depending on whether the society has near full employment or not, but in any case, the average earning that a person dying at midlife could otherwise have derived from working an average number of hours through a full life would be one quantifiable component of the VSL. Assuming an average 20 years of work lost at the time of premature death, this would, in the Danish case study, correspond to a value of about 4 million DKK[§] for an average income earner, disregarding subsidy incomes (which society no longer have to pay after the person dies, cf. Figs 3.7, 3.8 and 3.10). It follows from this estimate, that the ExternE VSL comprises a substantial component of intangible values, for example attributed to the citizen's contribution to social coherence, and perhaps also values reflecting ethical reasoning. The US Environmental Protection Agency has proposed a value of US$7.4 million in 2006 for a VSL in the United States (US EPA, 2015; Viscusi and Aldi, 2003), and notes the large differences between valuations in different parts of the world.

The valuation of VSL based on the loss of a member of society at midlife would indicate that the total stock value of an entire human life, which to distinguish it from VSL may be termed SVL, the *statistical value of a life*, is about twice, that is the capital asset associated with a healthy member of society in the European framework is of the order of $9 million. The human asset provided by the Danish population of 5.6 million would thus be about $50,000 \times 10^9$ 2010-DKK, from which the educational capital of $29,000 \times 10^9$ 2010-DKK (Fig. 3.10) already counted should be subtracted. The resulting health asset would thus be about $21,000 \times 10^9$ 2010-DKK (about $3,000 \times 10^9$ US$/cap.). Recent variations in the two components are included and shown in Fig. 3.10. In the context of the components of human capital discussed at the outset of Section 3.1.3, the educational component has been sorted out and estimated in isolation, whereas the life expectancy and SVL-based component combines health, educational and cultural contributions to the human capital. Subtracting the part connected with work ability, the remaining part is determined by the chosen magnitude of the SVL and further division specifying the role of culture versus general social coherence cannot be made. As mentioned earlier, significant contributions to literature, art, music and science are often beyond valuation, which in economic terms translates into a human heritage of

---

[§]1 DKK is about €0.13, or US$ 0.15 by mid-2015, but was US $0.20 in 1995.

a value that must be taken as infinite (putting in perspective political or religious extremists purposely destroying such assets).

3.1.3.3 Adding Up Wealth

Returning to the list of human assets or wealth listed in the beginning of Section 3.1.3, most of the natural assets (all types of land, sea and natural biota) should be accorded infinite value. As regards subsoil resources, depletable resources can be given a value, as it was done in the Danish case study for fossil fuels (Fig. 3.3), whereas others, such as clay, sand, gravel and stone, have a near-infinite value, partly because they may be reused and partly because—at least in the Danish context—they are renewed by rock withering and decay elsewhere, followed by sea transport to Danish shores. The produced assets are in principle much easier to list and valuate. The dominant component is buildings, as shown in Fig. 3.4. Net financial assets, comprising monetary instruments not already counted as representing physical assets (ie most stocks and bonds) minus debts, should be considered as assets. For Denmark, the net value is modest (Fig. 3.3). Finally, the human assets estimated in Fig. 3.10 round off the picture. They are considerably larger than the produced assets, which again are larger than the remaining valued components of wealth, but still one should keep in mind the infinite value of assets such as land, which have dropped outside the margins of these pages.

3.1.3.4 Inequality

Inequality in income has been monitored since the first publications around 1900 from Statistics Denmark and from similar agencies in other countries, sometimes condensing the data into a single number such as the Gini coefficient (Gini, 1912; partial English translation by Ceriani and Verme, 2012), and sometimes focusing on the disparate concentration of income in a small group of wealthy individuals within a country (e.g., Piketty and Saez, 2013, 2014), with particular focus on long time series of variations in the income share of such minorities groups. In the United States, the top earning 10% currently presently has a larger share in income than ever before, and in Europe one that has not been seen since 1939. Figs 3.11–3.13 show the family income distribution for Denmark in 2012, displayed in three different ways. The first just lists how many families have earnings in a given monetary interval, the second gives the share of total income for a certain fraction of the families, ordered from low to high earnings and the third arranges this material in terms of 10 equal-size family groups (deciles). Families are defined as groups of people living together in a household, including single individuals as well as couples with or without children, and income is derived from tax authority information, including all types of income, including government subsidies, but all before tax.

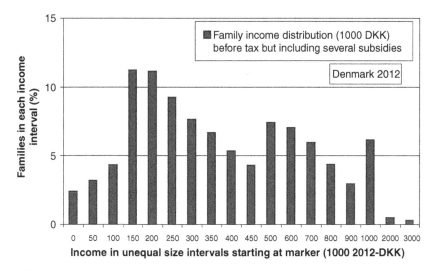

FIGURE 3.11   Distribution of Danish family incomes 2012, without subtraction of income tax but including both salaries, business surpluses, interest earned and income transferred from pension funds, insurances and public subsidy transfers. The discontinuities at 500,000 and 1,000,000 DKK are mainly due to the change of scale (Statistics Denmark, 2014).

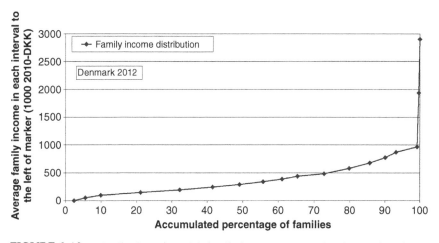

FIGURE 3.12   Distribution of Danish family incomes 2012, as in Fig. 3.11 but shown according to accumulated number of families instead of size of income. The data points are again not equidistant (Statistics Denmark, 2014).

In view of the Danish introduction of the welfare economic paradigm during the 1930s, followed up after World War II and continually expanding until about 1990, where the neo-liberalistic paradigm took over, it is instructive to take a look at the changes in Danish income and wealth inequality over the 20th century. Fig. 3.14 looks at the income distributions

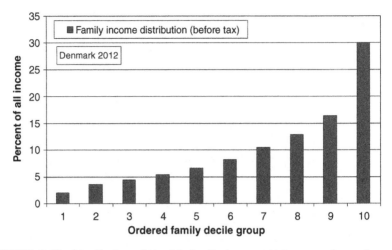

FIGURE 3.13 Distribution of Danish family incomes 2012, now presented as percentages of all incomes ordered according to equal family number segments (Statistics Denmark, 2014).

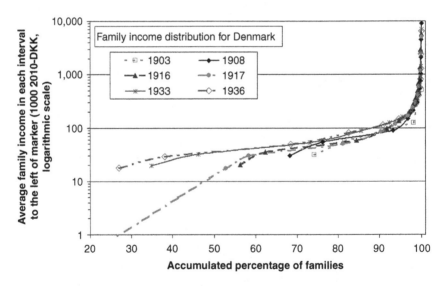

FIGURE 3.14 Family income distribution for Denmark for selected years between 1903 and 1936, shown according to accumulated percentage of families as in Fig. 3.12, but in fixed 2010 Danish Kroner (1 2010-DKK is roughly 0.2 2010-US $). The statistical data collection did not during the period in question consider incomes below 800 DKK (around 40,000 2010-DKK), but Statistics Denmark did try to extrapolate the curve for 1917 downward, using the known number of families with no reported income at all, around 7%. The definition of a family is discussed in the text (Statistics Denmark, 1919, 1938).

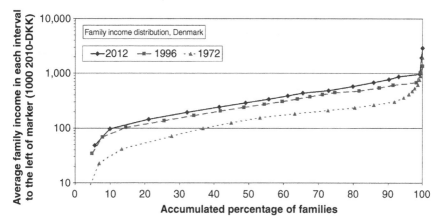

FIGURE 3.15 Family income distribution for Denmark for selected years between 1972 and 2012, shown according to accumulated percentage of families as in Fig. 3.12, but in fixed 2010 Danish Kroner (1 2010-DKK is roughly 0.2 2010-US $). The definition of a family is discussed in the text (Statistics Denmark, 1975, 1998, 2014).

during the two periods 1903–17 before the welfare economy introduction and 1933–36 during the actual introduction effort. Fig. 3.15 does the same for 1972 (at the end of an exceptional economic growth period), for 1996 (before the neo-liberal policies made much of an indent), as well as for 2012, where Denmark like other European countries is still struggling with the aftermath of the 2007 bubble economy crash. The early statistics (Fig. 3.14) did not employ the concept of a family but it also excluded married women and children below age 18 years, implying that the set of tax return respondents used by Statistics Denmark is fairly close to that designated as families in more recent statistics. The postwar practice (used for 1972 in Fig. 3.15) was to include independent taxpayers down to age 15 but to lump married couples and their children together, again effectively using a family or household concept and attributing most income to a 'head of household'. Later, legislation aimed to limit underpaid child work, and the change in educational enrolment (Fig. 3.6), made the income from people below18 years become less important. As noted, the recent statistics just define families as groups of one or more people living together.

The early statistics (Fig. 3.14) shows a very large dispersal of incomes (note the logarithmic scale), particularly for the highest incomes, earned by the top 10 and top 2% of the accumulated families, where incomes are 10–100 times higher than at the 90% mark. The top-end inequality is diminished in the more recent data (Fig. 3.15), typically to less than a factor of five above the income at the 90% mark. Even the recent neo-liberal mantra of encouraging income differences has only produced a moderate increase in inequality, notably in the accumulation interval 80–98%, and still much less than for the period 1903–08. The achievements of the welfare economy are very clear in the 1972 data and persist to the 1996 situation. The entire curve is shifted upward except for the 92–98% accumulated family percentage interval, where the rise disappears, and furthermore, the percentage increase in income is consistently larger the lower the income is, down to the lowest incomes. The income redistribution policy of the welfare period has in this sense worked well up to 1996, but its original intent to reduce the disparity between the richest and the poorest to under a factor of three has certainly not been realized. The absolute income level in fixed prices indicated by the middle part of the distribution curve was largely unchanged between 1903 and 1936, but by 1972 has increased by more than a factor of three, and from 1972 to 2012 has further doubled. The reasons for the overall increase in income may well be global and local advances in (education-based) technology, but the advances with regard to equality are evidently caused by the redistribution policies of the welfare economy. The redistribution transfers significantly influence the incomes of the poorest and to some extent the richest families (as further discussed in connection with Fig. 3.16), and had redistributed income not been included in the figures, the curves would have been much more similar to those for 1903.

The early period depicted in Fig. 3.14 shows that the 1% of very rich families persists nearly unchanged from 1903 to 1938, but the income

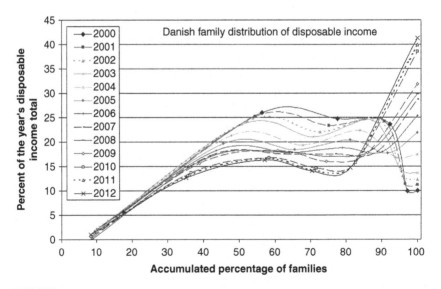

FIGURE 3.16 Recent developments in the family distribution of disposable income in Denmark (based on Statistics Denmark, 2015f). Disposable income is income after tax and interest payment, with a fictive value of housing rental added for house owners. The position of data points is moving between years, because Statistics Denmark uses the same income intervals for all years.

in the lower parts of the accumulated family distribution increases systematically, gradually diminishing the number of families at the lowest monitored income. Another change is a modest increase in income for the families in the accumulated 80–90% area, between 1908 and 1933. The deflator used to transform all incomes to the 2010 price level is the GNP-deflator after 1966 but the general consumer price index earlier. During the period where both are available, the difference between the two methods has been checked and found small enough to rarely make any noticeable change in the graphs.

One part of income redistribution is achieved by public subsidies, primarily to families of lower income, and it has been included in Figs 3.11–3.15. These transfer incomes amount to as much as half the income among citizens with only basic education, according to Figs 3.7 and 3.8. However, because the incomes are shown before tax, it is not disclosed where the subsidy money has come from. The welfare economy did base the transfers on a progressive tax rate, ensuring that the subsidies to the low-income groups were predominantly taken from the highest-income groups. Available data on taxation are for many of the early periods not ordered in the same way as income before tax and the available tax revenue data do not directly reflect redistribution of income because the tax money is used for many other purposes, including, for example infrastructure, schools and hospitals. In recent years, Statistics Denmark has tried to make the effect of redistribution stand out more clearly, by providing tables on a quantity called *disposable income*. It includes all income including subsidies, but subtracts income taxes (but not taxes on consumer goods), and it tries to compensate for the different monetary options of people owning their dwelling and those renting it. This is done by adding a fictive rental value for owned dwellings (a value already estimated by the tax authorities using it in connection with property tax calculations) and by further deducting the interest payments on real estate loans. However, as these are not separately known, instead all interest payments are deducted, for house owners as well as for renters. Fig. 3.16 shows the development of the Danish disposable family income from year 2000–12. There is an overall income growth in this period, but the curves have been normalized to the total disposable income for each year. The results indicate quite strikingly the transition from the preceding 70 years of welfare economy to a neo-liberal paradigm of rewarding the highest income earners. The top 3% of families by the year 2000 had a low share compared with the upper middle incomes, but by 2012 it has grown by a factor of four and much above that of the middle incomes. The middle-income groups are the losers in this development, while the low-income families are neither losing nor improving their shares of disposable income.

Inequality may be better described by wealth rather than running income. The distribution of assets provides a different and, in a sense, clearer display of the overall disparity between groups in a society or between different societies. On the other hand, data availability is often inferior to that of income, which most countries regularly monitor for taxation purposes. Several countries have abandoned general wealth tax (although still taxing possession of certain assets, such as land, buildings or stocks), so data series tend to be discontinuous and incomplete. In some countries, it has been possible to derive wealth information from assessments of death estates (Piketty et al., 2006).

In the case of Denmark, some data are generally available up to 1996 because tax returns had to specify wealth, but this disappeared from 1997 when wealth taxation was abandoned. Later accounts only cover real estate and stock holdings, toward which taxes were still levied, and most of the wealth connected with those individuals having a personal business is no longer recorded. However, also in 1996 and earlier, the statistics are incomplete. Funds bound in pension and life insurance schemes for later payments were not included (but estimates such as the one in Fig. 3.3 are available) and neither were equipment, furniture, art or antiquities in family possession. Figs 3.17 and 3.18 gives the estimated distributions of Danish wealth during the early and late 20th century, plus a partial estimate for 2005.

The curves for selected years 1903–36 in Fig. 3.17 are remarkably similar for the most wealthy 4% of the families. Although wealth data below 400 2010-DKK were not collected in 1903, the slope of the curve indicates that little wealth was bestowed on the lower 85% of the families. By 1917, the curve reaches down to the 80% ordered and accumulated level, and

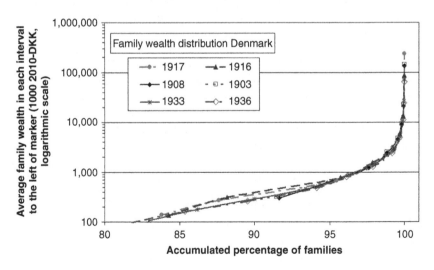

FIGURE 3.17   Distribution of Danish family wealth in selected years from 1903 to 1936 (Statistics Denmark, 1919, 1938).

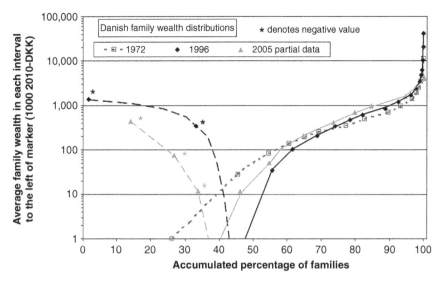

FIGURE 3.18 Distribution of Danish family wealth in selected years from 1972 to 2005 (Statistics Denmark, 1975, 1998, 2007; note negative values for 1996 and 2005).

although the top wealth is not significantly changed, the 85–96% accumulated family interval enjoys increased wealth. A major reason was the 'new-rich' strata of shady Danish businessmen, who profited from selling subquality canned food to consumers in the World War I-fighting European countries. After the war, many such businesses went bankrupt and took a large Danish bank down with them (Sørensen, 2012). The curves for the 1930s are much more similar to the 1903–08 curves than to those of the intermediary period, except that the peak wealth has dropped from above 100 million 2010-DKK to 30–60 million 2010-DKK. The explanation is the financial crises of 1930, which as seen took at least the wealth part of the Danish economy 30 years back.

Moving to the post–World War II period in Fig. 3.18, the 80–90% family interval in 1972 is not much changed relative to the prewar data, but the top wealth is reduced and the trend of increasing wealth extended to the lower parts of the distribution is continued. By 1996, the 72–98% family interval has gained wealth, the top wealth is a bit reduced, but first of all the wealth for the entire lower 72% of families is dramatically lowered. The data for 1996 and the partial data for 2005 show a new phenomenon of substantial negative wealth. This has become possible due to increased options for creating debt. By 1972 the Danish real estate financing institutions could only offer a loan of up to 80% of the estimated value of an owned dwelling (with associated land). Banks could provide further loans, but were very restrictive if no collateral security could be offered. People living in rented apartments had even fewer options for obtaining loans. The energy supply crisis in 1973/74 brought about the decision to allow loans in excess of the 80% limit for energy related building improvements, and during the 1990s, loans were generally offered without much consideration of collateral security, only with an appraisal of annual income. This continued into the following century, helped by the conviction that full employment was permanent and by a 2002 government decision to reduce real estate taxation, which caused house prices to skyrocket and gave the impression that house owners had plenty of collateral. One reason behind the old 80% limit had been a belief that house prices could fluctuate over time, but this seemed forgotten after year 2000, until the housing bubble burst in 2007 and brought house prices down by at least 12%.** However, Statistics Denmark has found that nearly half the negative wealth segment in Fig. 3.18 represents families not owning their dwelling. In both the house-owning and nonowning section, the financial crisis that began in 2007 created many bad-loan cases, due to loss of job or other loss of income, and several major Danish banks would have become insolvent had the government not decided to bail them out. Nationalizing the banking sector was not discussed and no conditions were set for the banks' future handling of loans. The banks responded by doubling the management salaries, seemingly using a neo-liberal style argument that the banks might have acted less foolishly if the bank management had been better paid. In any case, Fig. 3.18 shows that 45% of Danish families had a negative wealth in 1996, and 39% in 2005. The omitted contributions to the 2005 wealth data imply that the data points should be moved upward, so a comparison with earlier years is not meaningful. For all the years considered, the wealth disparity has been similar to or larger than the disparity in incomes.

3.1.4 Buying and Selling Goods

The current setup for trading at the consumer level involves a commercial sector acquiring materials or other goods from wholesale distributors, import businesses or directly from manufacturers, and selling them to individuals through a web of retail outlets, physically strategically located or selling through Internet sites with postal or parcel services taking care of delivery. Central in the approach to buyers is massive advertising through a variety of media, with advertising costs typically in the range of 5–50% of the sale price. Goods may be categorized as daily needs, such as primarily food and hygiene items, intermediate duration merchandise such as clothes, batteries, garden plants or toys and more durable consumer items such as computers, bicycles, refrigerators and cars. Typical

---

**For houses actually sold. In addition, many houses have, for years, been unsuccessfully for sale.

lifetimes for the three roughly defined categories are weeks, months to years and 15–25 years. This raises the question of what type of insurance the manufacturer and seller should be obliged to offer, as regards quality, functionality during the expected lifetime and provisions for repair and end-of-life handling.

Current legislation differs from country to country, but usually includes a codex for sober information before purchase, a declaration of the ingredients of the item sold, and a warranty typically much shorter than the expected product life. A few manufacturers offer better than legally required guarantees, such as 5–10 years for computer hard disks, automobiles and furniture. As regards 'sober information', most advertisements leave much to be desired, giving no serious information at all or giving misleading information by focusing on positive properties of the product in question but not mentioning areas where the product falls short of other products serving the same purpose. Certain software companies have tried to avoid the consumer legislation valid in some countries by stating that they do not sell their products but only lease them and therefore are not bound by consumer legislation (because consumer legislation in most countries do not deal with leasing, assumed to be of interest only for arrangements between companies).

From a resource and energy perspective, the current situation also leaves much to be desired. Unnecessary short lifetime of many products, as well as poor design with respect to ease of disassembly for reuse and recycling entails a greater-than-necessary resource and energy usage, forcing consumers to buy several consecutive products for the same purpose over a time period where one decent product would have sufficed. Not designing the products for reuse means that disposal options fall back to incineration, which usually limits recycling to certain metals. A slight concern for the environmental issues would lead to the requirements that customers are fully informed not only about the lifetime of a given product, but also is told where it stands compared with optimal products serving the same purpose. More serious concern would lead to a requirement of specific legislation governing lifetime, resource and energy use by the manufacturer and user, as well as a warranty obligation extending throughout the lifetime of the product under specified conditions of use. Isolated legislation along such lines is in place in some countries, for example for building shell–heat losses of new buildings, and in other countries it is approached by providing an open rating system for voluntary consideration at consumer purchase decisions. In Europe, this has, for example led to several classes of refrigerators no longer being produced because consumers always bought those in the top one or two energy rated classes. For cars, this rating system has led to a clear division of consumers into two groups, with one going for the best energy efficiency cars on the market, while the other group intentionally buys the least efficient vehicles, in order to exhibit their indifference to resource and environmental concerns.

A progressive list of measures for improving the situation would start with expanding the rules for declaring ingredients used in the product. The present full declarations required for food products should be used for all products that may be associated with human intake, such as lotions and other cosmetics applied to human skin and nearly all being internalized by absorption through the skin. The explosive rate of adding chemical substances to products that may consequently be harmful to humans would indicate the advantage of working with a fixed 'positive list' of allowed substances having undergone long-term testing. Consumers should also know the expected lifetime and the environmental and energy consumption characteristics of each product before purchase, meaning that all products offered for sale to consumers should have undergone a certified life-cycle analysis.

It would seem that current advertising is a poor way of communicating between sellers and buyers. Instead of praising one product over similar ones, the customers would be much better advised by having access to a searchable database where different offerings for each kind of product can be compared on an objective basis, as regards all the concerns mentioned earlier. Some such databases are already available on the Internet for a number of product groups, but they are commercial and would need certification from an independent overseer, such as a government committee operating in full openness. The manufacturers would of course have to pay a standard fee for having their products included in the database, but this would normally be considerable less than the current advertising costs. As it is not in any society's interest that people buy products they do not need or flawed products, a general prohibition of all advertising would contribute to heightening the quality of commerce and in consequence the welfare of consumers. An overseeing institution can handle disputes over the correctness of database information offered. Prices may go down as the advertising costs are avoided, and eliminating the dishonest manufacturers and sellers is also a positive move.

The end-of-life handling of products is another area where legislation may be necessary. In Germany a lively debate took place during the 1990s, suggesting that manufacturers reduced the number of components in their products and designed them to be easy to disassemble for reuse or recycling of components, and that manufacturers should be compelled to take the decommissioned product back for reuse/recycling after its service life. Several industrial companies expressed their sympathy with the idea, noting that, for example a car door in the past contained several hundred components, but that a better design would greatly diminish the number of components and make it easier to treat after its service life, particularly for the company that had manufactured the product in the first place. Some industries went along this way, for example by standardizing components so that all cars manufactured by a given company had the same door handles and other components that offered themselves to standardization across model variants. Unfortunately, this debate never went further, indicating that such requirements should rather be made compulsory, so that they impose the same requirements for all manufacturers and thus do not affect competitiveness among industries that know their business.

Legislative requirements of this type should not discourage small manufacturers. This would be avoided if insurance type schemes were available to rescue industries not having an internal economic capacity to absorb occasional mistakes.

Provided that leasing or rental access to consumer products are subjected to the same rules as purchased products, many consumer products (and in particular those of longer lifetimes) could be offered for rental rather than ownership, making it even more natural that the manufacturer assumes the responsibility to ensure the proper functioning of the product through its working life as well as in the disposal phase by taking the product back for proper treatment or disposal. If computer software manufacturers continue to insist that they only lease their products to consumers, it is even more important that the consumer protection legislation is the same for leasing and purchase, so that consumers are not buying deeply flawed software that the manufacturer expects to improve by frequent updates[tt] as the consumers identify the flaws.

This would also eliminate the discussions of guarantee periods, because warranties would effectively be for the entire product life. If purchase and change of ownership is the preferred way of handling commerce, the warranty rules will have to be changed in such a way that guarantees against manufacturing flaws prevail throughout the product life, while maintenance and wear during use is accepted by the customer, but on conditions that the need for and extent of these have been clearly communicated before purchase.

Following these simple rules will make significant contributions to improved resource and energy usage and will at most affect the general welfare in a positive fashion. Losers are in principle only businesses purposely wanting to cheat the people who buy their products. However, the suggestions will also affect the current practice of having consumer items manufactured in developing countries with low salaries, claiming that in exchange for a poorer quality and low product lifetime, one helps these countries develop. This is false for two reasons. First of all,

---

[tt]One manufacturer postulates a need to update daily, claiming that the main reason is new malware threats. In reality, it is difficult to prevent others from stealing customer information while allowing the software company themselves to collect any information they see fit and sell it to other commercial or intelligence enterprises.

intercontinental shipping of consumer products appears feasible only due to the current low cost of oil for (mainly ocean) transport,[‡‡] and secondly, only false advertising can influence consumers to think that buying five successive short-lived products made in an underdeveloped country is cheaper than buying one properly crafted product with the best lifetime possible and thus the smallest resource use, environmental impact and energy use. These remarks do not mean that international assistance cannot help to rapidly develop less developed regions. The requirements are to improve local education and offer initial design assistance from advanced countries, a method that has produced successful results notably in China, but along with other local product manufacture characterized by all the negative lifetime and quality issues mentioned.

3.1.5 Work Versus Meaningful Activities

The concept of work plays a large role in present societies because nearly all components of welfare have been monetized and are therefore only accessible to people who possess money. As seen from the case study in Section 3.1.3, very few people today have access to money unless they have a paid job or a personal business. For this reason work is considered essential and most people will accept much to keep this source of income. That has not always been so. Work is really a fairly new invention, introduced in connection with the Scottish industrialization during the 18th century (Anthony, 1977). Of course, toil had been common long before that, in order to provide the necessities of life or build defence walls as slaves for some king or emperor, but the philosophy of labor changed. Once, village societies had provided the necessary food and assets by just a few hours of daily activity, integrated into a meaningful life with time for leisure and enjoyment (Lee, 1979; Wiessner, 2002). Early city states, for example in Mesopotamia or Egypt, had organized division of tasks between classes of citizens, including forced labor, and similar arrangements are found all the way through history, to Medieval, Renaissance and Enlightening Periods. However, the introduction of factory work required new types of coercion because instead of persons determining the pace of labor activities they were now being dictated by the machines. Because investment in machines was seen as high compared with worker's salaries, pauses were no longer allowed and work in shifts was introduced. In order to help gaining acceptance for this new order of labor, the Protestant Church was very helpful by declaring that 'idleness is a sin'. Much has changed since the early industrialization, and studies have been made concluding that continuous work is not optimal for productivity,

[‡‡]The World Trade Organization has compelled international transport to be exempt from the environmental taxes prevailing in individual countries.

just as many societies have reduced the number of working hours in order to leave more time for social life (and of course to create more demand for the products of industrial machinery). Still, there is disagreement between those who think that the cake (total output) has to grow in order that the workers can get a modest salary rise aimed to prevent rebellion, and those who favor a redistribution of wealth, in order to possibly remove the need for a constantly rising growth and its negative effects on resources and environment (Illich, 1978; Gemynthe, 1998; Sørensen, 2014, Chapter 12).

In a more concrete way, it is proposed that the time has come to replace 'work' by 'meaningful activities', or rather to return to it (focusing on those activities of the past that made sense and gave satisfaction, as opposed to slavery). The implication is that anyone should have both an understanding of why a given job has to be performed and an acceptance of being the right person for the job. This means that society should be clearer in defining what has to be done in the interest of common welfare, and that education should be offered to make sure that each citizen can appreciate the job he or she is asked to perform and can do it well. In a way, elements of this attitude to the activities by which we earn our living can be seen in many current societies. Many people perform their work 'con amore', being satisfied with what they contribute, not caring too much of how long time it takes to finish the job or if the salary/income is a bit higher or a bit lower than that of the person next door. The problem is that a large portion of current work serves purposes that do not contribute to common welfare.

One possibility is to get rid of the concept of 'working hours'. The transitions that have taken place in many societies over the recent decades have removed much factory work (that could be further made superfluous by increased use of robot technology, leaving only a few operator and supervision employees, similar to what is seen, e.g., in the electric utility sector). In fact, many jobs need not have the employee come to a particular workplace (with negative implication for transportation energy and time use), but could be performed from where the person happens to be. Some people today oppose 'home work' because they say that it blurs the distinction between work and leisure time. But this may well be desirable and something one should work toward, along with redefining work as meaningful activity and eliminating current work types no longer seen as meaningful. There are jobs today that cannot immediately be 'flexed", such as those of bus and train drivers, power line maintenance units, hospitals and other medical assistance. However, even some such activities might in the near future become much more flexible: vehicle systems may be (and some already are) driverless with use of techniques such as computerized collision avoidance systems, education need not take place in a classroom but can proceed via telecommunication through phone or computer media. Still, there will be medical and social job functions, plus some management and supervision jobs that need to have personnel available at all times, but the 'meaningful activity' paradigm also assumes that everybody understands that there are such needs and are willing to undertake them in a planned combination with leisure and other activities.

In any case, a gradual transition to a society with emphasis on meaningful activities will impose many conditions on the technology and social arrangements that will best facilitate the new structure. Current waste of time and resources on commuting can be largely avoided, as can many sales activities such as shopping, that to a large extent needs no supermarkets or retail shops, but only an optimized system for transportation of goods. Similarly, service jobs are not optimized by the current setup of traveling craftsmen, and building and other construction activities are often made in ways that increase work as well as cost more than needed. If such activities are conducted under a system of fixed price and quality guarantee, for example by creating a public company with optimal cost structure, then free competition from any number of private small or large companies can be accepted without risking ending in permanent high-cost markets such as is often the case now. Evidently, setting up such public/private competition is only one suggestion of how the present inflexibilities of purely private building and repair sectors can be avoided.

3.1.6 Tolerance and Multiple Solutions

The last remarks in Section 3.1.5 are examples of the principle of not necessarily choosing a single solution, but to allow different solutions to coexist side by side. The world has seen centralized communism and neoliberalism fail (the latter unfortunately unnoticed by quite many people), and any proposed 'third way' may well with time similarly lose its attraction. The idea of creating a system that allows different paradigms to live together in peace may be a better solution. It requires tolerance from each of the coexisting schemes, but this is precisely the quality that has to prevail in any functioning society, so it is not a bad thing to make it stand out as a fundamental condition. This is obvious if one looks at the causes for conflict in the past. Tolerance toward different views on religious matters, on social setup and interhuman interactions must be introduced and generally accepted if we are to avoid new conflicts based on fundamentalist views on how other people ought to behave. Tolerance must be part of our upbringing of children, which clearly means that parents cannot be allowed to transmit their prejudices to the children. This is one basic condition for improving the lot of our societies and will therefore have to be incorporated in the fundamental rules of human societies. How to approach that is the subject of the following sections.

3.2 HOW TO FORMULATE CONSTITUTIONS

Constitutions should deal with fundamental conditions for living in a society and leave debatable issues to daily political deliberations, restricting the constitutional text to setting the general frame for political debate and resolutions. The main text of a constitution would therefore be the formulation of basic human rights and duties in the context of the structure and maturity of the society in question, as well as defining the type of governance that should be used to ensure compliance with and further development under the set rules. Section 2.3.1 gave the UN's bid for formulating a basic set of human rights, and Section 2.3.2 outlined governance schemes currently employed. This chapter will discuss ways that have been suggested as well as new ways of tackling the problem of governing a society, and will also question if the division of societies into national states is the best way of achieving societies that function in accordance with the human rights and duties accepted.

Some current constitutions go further than stating human rights and frameworks for making political decisions and ensuring that citizens comply with both, in that they also favor specific economic, religious or other social paradigms. Examples of this are both in constitutions and the UN set of human rights that include 'property rights' and 'inheritance rights', and also 'enrolment in a specific religious institution' (Section 2.3.1, in the list of basic human rights and duties). By doing this, the constitution of course loses some of its fundamental standing and become a vehicle for conflict, inside the society as well as in relation to other societies with possibly different paradigms in their constitutions. Stating basic human rights in a constitution has therefore historically not prevented ferocious wars and crusades aimed to remove those thinking differently from the surface of the Earth.

Before going systematically into the range of issues involved in writing alternative constitutions, a few remarks will be made on the role of economic paradigms in defining the rules by which societies operate. Some of these have already been introduced in Chapter 2. On one side there are the neo-liberalists, who claim that although the conditions shown by Adam Smith to have to be fulfilled for classical liberalism to work are far from fulfilled in any current society, the 'market knows best' paradigm may still work, and at least it helps increase inequality, which they believe is a desirable trait because it favors indiscriminate economic growth as defined by the GNP. On the other side stands the remains of the socialist movement, which has lost lots of appeal after the successful media campaign staged by the liberals, claiming that the fall of the centralized oligarchic rule in the Soviet Union was the end of socialism. It may be instructive to attempt an objective assessment of how these two paradigms stand in current national societies.

As said, neo-liberalism aims to maximize inequality. Studies such as the Danish case study described in connection with Figs 3.11–3.18 shows that inequality has indeed increased during the last 10–20 years, but is still much less than in the early 20th century. The 10–99% interval of the income-ordered families have been in a relatively stable situation for several decades, although with an education-related inequality of a factor of 10, but tax-financed redistribution of income has rendered the lot of the lower range of families tolerable. The neo-liberal politics that took over from the 1990s caused this to change from around 2005 (Fig. 3.16), but the relative loss by the middle incomes have been hidden for those involved by the overall growth in incomes (Fig. 3.15). The situation in several other countries have been similar, independent of whether they have gone through a period of 'welfare economy' or not, but the level of inequality at the top differs and is quite high for countries such as the United States (Ohlsson et al., 2006). In other words, many societies have developed in the direction of less social solidarity as regards equality but at least a desire to remove beggars from the gutter.

It must be realized that true neo-liberalists are not unhappy about the recurrent financial crises inherent in the capitalist paradigm. These offer great opportunities for the rich to make lots of additional money, for example by intelligent buying and selling of shares. It is only the 'common people' that suffer from destruction of welfare by each crisis.

Taking a look at the fate of the variant of socialism called 'communism', in its original form, one may turn to the manifest written by Marx and Engels (1848), which can be summarized as follows:

1. Abolish property rights to land.
2. Use progressive income taxation.
3. Abolish inheritance.
4. Allow only state banks to issue loans.
5. Strategic sectors such as communication and transportation should be operated by the state.
6. Make the state in charge of agricultural and industrial production.
7. Equal liability of labor (put agriculture and industry workers on the same footing).
8. Gradually decentralize settlements.
9. Abolish child labor.
10. Make primary education free and leave further education to industry.

The manifest contains blatant criticism of socialist movements outside Germany, such as the French with their preoccupation with literary beauty and Proudhon's acceptance of modern technology, as well as the British bourgeoisie criticism seen as restricted to administrative adjustments. As Marx and Engels see it, only armed revolution can put the proletariat in power. Poor treatment of women in factories is briefly touched upon, but there is no word on free medical care or other social security reforms advocated by the other socialists. To Marx and Engels there are only two classes: workers and bourgeoisie, and all other socialist ideas aim to eliminate workers by lifting them to bourgeoisie status. Marxism offers no explanation of why it is important to preserve a class of workers with only primary school education, and if these workers take over the means of production, it is difficult to see the possibility of any improvement in outdated and polluting technology, a fear that has been fully borne out by the Chinese 'Cultural Revolution'.

Looking at the points listed earlier, the second (progressive taxation) has gained universal acceptance, albeit with different severity, the strategic sector ideas in point 5 was pursued in several countries before the privatization wave of neoliberalism and reducing child work and making primary education free (points 9 and 10) have been accepted in large parts of the world. The other points have not been supported, except that state-operated industries and agriculture have been tried for a while in the Soviet Union and China but subsequently given up. The top-end income and wealth disparity mapped in the preceding section may remain a problem today, but the number of people in the extreme top-segment seems too small to warrant upholding the notion of a classical class society. If the distribution problem finds a solution, it is thus unlikely to be by revolutionary fights between 98% of the population and the last 2%.

One dogma shared by Marxism and capitalism is the faith in economic growth. The difference has been in choosing the distribution of proceeds between workers and owners or investors. In capitalist theory the production factors are monetary inputs (investments), means of production (tools, machinery and intellectual input including but not limited to patents) and labor input. They share the profit in some proportion, determined by the owners or open to negotiation in countries with bargaining systems implemented, and in recent times occasionally regulated by public law, such as by setting minimum wages. Investors/owners tend to have more influence than other production factors, and in particular much more than the essential input from the inventors of desirable products and suitable production processes. In Marxist theory, the owners are replaced by the state, and the investment in means of production (based on an overall plan made by the state) is not accorded any share of the profit at all, implying that any increase in profit derived from improvements of technology is supposed to be given to the workers (although they did not have the skills to contribute to the technological advance).[§§] If a robot can do the work of 100 human workers, the one worker operating it would in principle earn 100 times more profit (which, of course, did not happen in actual

[§§]For mathematical details, see Sørensen (2010, Chapter 7).

communist societies). Marx was wrong in giving the profits from new technology to the workers, but capitalism was equally wrong in giving them to investors that also make no important contribution, but just hope to make their wealth, often acquired by inheritance, grow further.

By the latter part of the 19th century, the international socialist movement was much more inclined to follow Mikhail Bakunin (1814–1876) than his opponent Karl Marx, caught in making false accusations accusing Bakunin of being an imperial agent (Wikipedia, 2015a). Bakunin (1882) was anarchist in the spirit of his friend Proudhon and he rejected all authorities, including churches, social classes and Marx's 'dictatorship of the proletariat', which he (correctly) predicted would soon become more authoritarian than the regime of the Russian Czar, in favor of total individual freedom. His idea of self-governing communes was implemented by the noncommunist but short-lived Paris Commune formed in 1871. The rejection of any hierarchical social organization may sound as relying on direct democracy, but Bakunin was never able to clarify who should see to the implementation of decisions taken by individuals on matters affecting more than themselves.

The revolutionary socialist parties formed in several countries around 1900 were challenged by the more moderate Social Democratic parties, notably the French 'Socialist Party' led by Jean Jaurès from 1902 and advocating a program formulated at a congress in Tours, comprising

1. General elections for parliament (all citizens can vote) based on lists of candidates (not parties)
2. Direct democracy (referenda) used for important decisions
3. Freedom of the press
4. Equal rights for women and men
5. Administrative freedom for local governments
6. Separation of church and state
7. Churches and other congregations may not interfere in politics
8. Legal council and representation should be free
9. Abolish the death penalty
10. Make all courts civil, no military courts

All of these points can at present be found in constitutions of many advanced democracies, and most them have been generally accepted as desirable anywhere. Jaurés was assassinated in 1914 for his antimilitary views, and the nationalization intents of the 1902 congress drowned in conflicts between French socialist fractions and the charismatic political rhetoric of Georges Clemenceau, representing the Radical Party (Orth, 1913; Wikipedia, 2015b). The specific welfare variety of the Social Democratic paradigm, which was subsequently developed in the Nordic countries, has already been described in Section 2.3.3, because it has been an actually implemented form of governance starting at the time of the

1929–30 economic crisis and still partially remaining, and with several elements copied to other, mostly European countries.

The following sections will discuss further ideas for remedying the problems in the current political paradigms, of which the inequality discussed in Section 3.1.3 is an essential part. A general remedy of course remains using progressive taxation of income and wealth, combined with assistance to low-income groups, as discussed in the Section 3.1.3 case study in terms of redistribution, but even the fairly aggressive redistribution of the early Danish welfare economy is seen not to remove the advantage of the top-incomes and top-wealth segments, due to the Social Democratic promise not to interfere with the private sector entrepreneurship in general. One must therefore be wary of proposals to solve the inequality problem by taxation without fundamental change in the underlying economic paradigm (e.g., Piketty, 2014).

3.2.1 From Ancient Greece to Revolutionary France and Liberal USA

The rulers of the Greek city states were quite aware, in 7th century BC,*** that states could be ruled by one person, by a small group of persons or by many persons, and writing down the preferred rules in constitutions was not at all foreign to them (including the first known practice of voting; Larsen, 1949). What happened over the following two to three centuries was that more and more governance concepts resembling those we use today were discussed and sometimes introduced. The first significant step forward was made by Solon (638–558 BC) in the city state of Athens (including the surroundings of the city). He forbade using children as collateral for loans, freed existing slaves and extended the previous administration by noblemen selected among themselves to comprise all (male) citizens, now eligible for an assembly responsible for legislation (but still based on proposals by the much smaller council of upper class citizens), and for justice through a new court where any male citizen could be called to sit in a jury. Lottery-style selection for office was used. Surviving poems by Solon exhibit a humanistic inclination, expressing contempt for the greed of many Athenians (Hansen, 1989). Legislation was introduced to limit dowries and inheritance, but after Solon's retirement, many of his reforms were abandoned, eventually leading to new dictatorial rule by a 'tyrant'.

After the collapse of the reign of a sequence of tyrants, two men wanted to rule Athens. The initially preferred candidate exiled hundreds of citizens in order to rob them of their property, and a revolt by the city and its council ousted him and recalled the second candidate, Cleisthenes,

---

***BC, before the currently agreed zero, a little over 2000 years ago.

who in 508–507 BC reorganized the various assemblies and courts in order to make them better reflect the composition of the city's inhabitants. In particular, the basic constituencies (tribes) were redefined to better reflect common interests and avoid quarrels. Proposing new laws remained the task of the 'upper house' council, as under Solon, but the main assembly, not formed by elections but by willingness to appear, convened up to 40 times a year and could pass as well as reject proposed legislation. The lottery was still in use for selecting administrators, but in case of the council partially ceded to elections by the assembly (Aristotle, 328 BC). Cleisthenes may have increased equality before the law, but he also allowed an assembly majority to exile citizens they did not like, with very vague charges.

Solon and Cleisthenes were rulers who wanted to consider the opinions of a wider part of the population, through staged assemblies for the different strata of society. Similar 'enlightened' autocrats are known throughout history in all parts of the world. Actually passing concrete power to the Athenian people still lacked an underlying framework (such as a declaration of human rights), and providing the assembly with the direct power to exile enemies is a clear example of the dangers of ill-prepared approaches to rights and duties in a democracy.

The next step in Athens' development toward a direct democracy was taken under the leadership of Pericles (495–429 BC). He furthered the ongoing process of strengthening the power of the courts over the assembly, and removed much of the remaining power vested with the council, with help from his friend Ephialtes. He had no reservation to using the existing law to have the assembly exile two of his worst political opponent. The ruling 'office of the strategos' had 10 elected members of which he was only one. Thus he ruled as long as he was looked up to and things changed toward the end of his life. Pericles' views are primarily known through a speech he made in the winter 431/430 for the victims of the Peloponnesian War, recorded by the historian Thucydides (431 BC). Pericles expresses the view that Athens is the most advanced democracy of its time, expressing its merits in this way.

> Our constitution does not copy the laws of neighboring states; we are rather a pattern to others than imitators ourselves. Its administration favors the many instead of the few; this is why it is called a democracy. If we look to the laws, they afford equal justice to all in their private differences; if no social standing, advancement in public life falls to reputation for capacity, class considerations not being allowed to interfere with merit; nor again does poverty bar the way, if a man is able to serve the state, he is not hindered by the obscurity of his condition. The freedom, which we enjoy in our government, extends also to our ordinary life. There, far from exercising a jealous surveillance over each other, we do not feel called upon to be angry with our neighbor for doing what he likes, or even to indulge in those injurious looks, which cannot fail to be offensive, although they inflict no positive penalty. But all this ease in our private relations does not make us lawless as citizens. Against this fear is our chief safeguard, teaching us to obey the magistrates and the laws, particularly such

as regard the protection of the injured, whether they are actually on the statute book, or belong to that code which, although unwritten, yet cannot be broken without acknowledged disgrace.

The part of the speech claiming that Athenians could treat their differences in an urban way does not seem to agree with actual events, such as the exiling manoeuvres and perhaps the fact that the assembly on more than one occasion tried Pericles himself under a variety of accusations and fined him substantial sums. He died in a devastating plague epidemic, and left an Athens subject to increasingly errant governance (Aristotle, 328 BC).

The dominating role of the assembly increasingly took the form of mob rule. A failed assault on Sicily and new attacks from another city state, Sparta, added to destabilization. In 406 BC 10 generals (including Pericles' son) managed to save Athens and its grain import by winning the naval battle at Arginusae. However, on return they were sentenced to death and executed by the assembly, on a charge of not having picked up survivors in the water. The assembly was here in violation of three provisions of the constitution: the judging was made by the assembly without involving the courts, the generals were sentenced collectively and not individually as required by the constitution and the accused were not allowed to defend themselves. Furthermore, attempts by citizens to use their right to question the decision of the assembly were rejected with threats of death penalty. A similar death sentence by the assembly befell the philosopher Socrates in 399 BC (Lewis, 2011).

These events had a profound influence on the philosopher Plato's view on the Athenian democracy. Plato (c.424–347 BC) saw democracy as the worst kind of governance, which could only lead to collapse and take-over by a tyrant (dictator). Instead, he suggested that governments should be led by philosophers (such as himself), and that the administrators (guardians) of the state were free to lie and break the law as long as it benefitted the state (current politicians would have loved him!). On the other hand, they should not own personal property and they should live in forced marriages and have their children removed to some remote countryside (an idea later reused by Mao in China). The general population exists only to serve the state. He uses an elegant, albeit increasingly monotonous, literary style, in which he puts his words in the mouth of the late Socrates and his friends, assembling regularly to get drunk and discuss politics (Plato, 380 BC, 345 BC; Hansen, 2013a).

After the massive failures of Athens, militarily and constitutionally, the democracy is revived in a milder form by introducing a number of restrictions, notably restoring the role of the judiciary system and giving the council a little more to say (Hansen, 1999). Also around 400 BC, the assembly attendance is rewarded by a pay, like the one Pericles had introduced earlier (about 450 BC) for jurors, administrators and council members. The public service payment was financed, first by voluntary contributions from wealthy citizens, and later by taxing all persons possessing wealth. This created an income redistribution contributing positively to furthering equality in the city state (Lyttkens, 1994, 2010).

Plato's pupil Aristotle (384–322 BC) observed the 3rd century BC adaptation process of Athenian democracy and, in contrast to his teacher, concluded that democracy could be made to work and in that case would be one of the best choices of constitution, at least for the city state that was his prime interest. He does not exclude the possibility of gaining stability by mixing oligarchy and democracy, and like Plato he finds it necessary that children receive a uniform public education, rather than being influenced by their parents. To define 'best choice', he used a measuring stick of welfare, which he denoted 'happiness' (Aristotle, 326 BC; Hansen, 2013b). Aristotle was not only a philosopher but also a broadly oriented interdisciplinary scientist, making contributions to several different fields. As a general comment to Aristotle's *Politics* and the Athenian constitutions, Rubin (2001) notes that Aristotle is very clear on stating that electing rulers and legislators by lottery and limiting the time a person can sit in office is democracy, but representative elections by vote is oligarchy. The city state Sparta (in perpetual conflict with Athens), where all offices are filled by election is thus an oligarchy, despite its egalitarian law that all citizens should wear similar clothes and eat the same food, and so would any present nation be, despite our flirting with the name democracy.

After the Athenian democracy is nullified following another military defeat in 322 BC, not much was heard about democracy until the 13th century, and even then only sporadically, in connection with minor autonomous areas in Europe. An exception is the Roman writer Cicero's reiteration of the constitutional questions raised by Plato and Aristotle. Cicero (54 BC) comes out in preference of monarchy but is willing to accept some features of other governance forms. The actual constitutional basis of the Roman Republic was first a kind of direct democracy (by popular assemblies and more restricted councils), which elected magistrates running the passing of legislation, the courts and the military activities. Later, the imperial Rome was ruled by dictators conferring with various representative bodies (Wikipedia, 2015c). The 'things' of the North European Viking period around 900 were probably no more than councils in which the leader conferred with or was elected by important landowners and warlords (Sørensen, 2011). William of Moerbeke's translation to Latin of Aristotle's *Politics* appeared around 1260 (Bolgar, 1958) and in 1291, the three autonomous Swiss regions (cantons) Schwyz, Unterwalden and Uri instituted a federation with elements of direct democracy, which later developed into the present Swiss Federation. Already before signing the 1291 treaty, the three forest states had instituted community management of forests, pastures and livestock herds, based on frequent assemblies (Landsgemeinde) of all interested male inhabitants, agreeing on actions on these issues by simple majority vote. This might be interpreted as bringing out the suitability of direct democracy for smaller communities such as cities or here individual, modest-size regions. Already the federation-forming treaty of the three Swiss regions was decided not by direct democracy but by representatives elected by the three individual assemblies (Elazar, 1993).

Over the following centuries, several city states appeared in Europe (such as Andorra, Monaco, San Marino and Liechtenstein), with Republican-style government and various degrees of direct democracy. One feature was the introduction of a legal system with judges elected by the assemblies, and charging these with an arbitration role in case of conflicts with other regions or states, and with functioning as courts for criminal charges within the region. The Swiss cantons were during the Renaissance period challenged by the Church, and later by Napoleon during his imperialistic adventures, but the federal structure as well as many of the local direct democracy traditions has, in fact, been preserved to the present.

Elazar (1993) described the difference between Swiss democracy and, for example US democracy by noting that in Switzerland, individual freedom is promoted inside the liberty frame provided by the community, not outside and not above it. Community democracy requires citizens to contribute to the community and to show tolerance toward others, including minorities, rather than worshipping individualism as the highest goal. One should remember that the success of South European city states in general during the medieval period was due to the difficulty of maintaining central control over large areas by the kings and other rulers following but much less organized than those of the Roman Empire. Most of these city states, such as the Italian cities of, for example Venice, Turin and Florence, had oligarchic or dictatorial rulers, although considerations of trade and commerce created powerful interest groups that sometimes were able to force a political debate. Further North, in Germany, the Lowlands and Scandinavia, cities were less important, but a similar fight emerged between local landlords and kings trying to establish larger territorial states, neither being keen on democratic participation in governance. In several countries, the kings involved the Roman Catholic Church in an attempt to exert a comprehensive mind control over the populations (Downing, 1989; Sørensen, 2012).

Town meetings in New England had some similarity to European city assemblies. In 1628 the British king authorized private companies such as the Massachusetts Bay Company to run the local government including courts and law making (Zimmerman, 1999, Chapter 2). Town meetings were at first spontaneous and without any formal basis, but attendance soon became compulsory and before the end of the 17th century, most town matters were indeed decided at these meetings. In 1634, the General

Court imposed on any male above 20 years of age living in the colony, that an oath be taken that he and his family would be obedient to the magistrate and obey any law enacted, and promise to report others who broke the laws, as codified in the 1636 *Pilgrim Code of Law* (Lutz, 1998, Chapter 20). Each state passed a similar code of law, often with constitutional statements of human rights, such as in the 1639 *Fundamental Orders of Connecticut*, the 1650 *Connecticut Code of Law* and the 1647 Rhode Island *Acts and Orders* (Chapters 52 and 39 in Lutz, 1998), but these were all aimed at securing law and order in the British crown colony and did not set rules for democratic activity, although the documents themselves were agreed at meetings and often after general voting. Human rights and liberties had been covered in documents such as the 1641 *Massachusetts Body of Liberties* (Chapter 22 in Lutz, 1998), partly based on the English *Magna Carta* (Anonymous, 1215), an unsuccessful attempt to set some basic rules for an armistice. These human rights and constitution precursor documents were heavily influenced by the church, which in the New England case meant by the puritan protestant persuasion. 'Puritans' was the common name given to a variety of church groups in Britain who fought against the autocracy of the king and wanted to constrain him from engaging in arbitrary wars, while establishing an level of moral-based personal freedom and liberty for the citizens (the debate is preserved in the *Putney debates* from 1647, see Clarke and Woodhouse, 1938; further discussion of their New England impact on the democracy debates may be found in Zimmerman, 1999; Chapters 3–8).

The 'modern' type of democracy originates from debates during the period of Enlightenment and gets implemented as a constitution by the French Revolution in 1791 and the founding US constitution, drafted in 1787 and coming into effect over the next couple of years. Thomas Jefferson and Benjamin Franklin visited France and participated in formulating the French *Rights of Man* declaration (Anonymous, 1789), which like the US *Bill of Rights* form the basis for the constitution text itself. This is the first time that a constitution is directly formed as a superstructure on a declaration of human rights, and at least in this sense does precisely what the Greek constitutions had failed to do. The modern version of democracy is representative and has no relation to the classical Greek form of direct democracy, and in contrast to the city democracies mentioned earlier was not similar to or influenced by the Greek experiences (Rubin, 2001; Markoff, 1999). Not all British were against the independence and republic formation in colonies such as America. Thomas Paine (1776) noted the absurdity of kings and hereditary succession as a prelude to his call to stop British barbarity and let the United States be free. An exhaustive and sober account of the US society and its political structure in the 1830s has been provided by an envoy of the French government, Alexis Tocqueville (1835, 1840).

The contents of the French and US constitutions are quite similar,[†††] except for a few amendments seen as required due to the particular situation of the United States after the Civil War, for instance allowing soldiers to keep their weapons and act as police in the western states, where no regular police force could be formed immediately. At least some of these amendments were supposed to be deleted later, and it is a gross misinterpretation of the permission for ex-soldiers to act as police, when conservative Americans currently claim that the constitution allows every man to bear firearms and use them, as it now happens down to elementary school quarrels. The 'right' to defend one's property with firearms is rooted in Locke (1689), who advocated (in contrast to Proudhon, see Section 2.1.1) that man should have full freedom to do as he pleased with any part of his property, whether physical or intellectual, including use of any means to defend it against intruders.

The intellectual debate on human rights and constitutions was initiated by Jean-Jacques Rousseau (1712–1778), along with his work in literature and music. In his analysis of inequality (Rousseau, 1755), he asserts that man shall be loyal only to the entire common good, meaning the welfare of the total society, rather than some monarch or oligarch. He also expresses his antiwar sentiment and disdain for private property, an issue that he follows up in his main political opus entitled *The Social Contract* (Rousseau, 1762). The contract binds man to serve other members of society in solidarity, and rejects the idea that each man has some 'natural rights', such as for the land he is able to be the first to lay hands upon. Rousseau's concept of a constitution distinguishes between laws, which he believes should be suggested and passed by direct democracy, and their administration by governments and courts, which should be populated by civil servants, hired or elected for their skills and knowledge. His treatise is not suggesting concrete paragraphs of constitutions, but keep the discussion on the level of principles, however not avoiding strong statements such as 'Representative democracy is one day of democracy followed by four years where the people are nothing but slaves'. Voltaire (1694–1778) criticized the inequality essay and called Rousseau a romantic 'back-to-nature' freak, favoring inarticulate and fighting mobs over proper government (e.g., in Candide; Voltaire, 1759), which might have been fair criticism had it been directed at Rousseau's early literary works and not the political works described here, where the responsibility toward the entire society is based on articulated demands and not uninformed quarrel. The attack on those preferring life to be as simple and as close to pristine nature as possible should rather have been directed at the American writer Thoreau (1754).

[†††]The New York Statue of Liberty was actually a gift from the French Government, acknowledging the adoption of the French Constitution rather than seeing it as a joint project.

3.2.2 Rights and Duties

The human rights debate following the increasing popular interest in these matters in Europe and its colonies after the Renaissance, and the constitutions subsequently passed in various countries, bore witness of an increased longing for personal freedom embracing a larger fraction of the populations than just the upper classes. Several constitutions, such as the British, emphasized liberty as a general good, while maintaining a firm restriction of governmental power to the king and his sworn allies in church and nobility. If rights were increasing, duties were still confined to obeying the laws of the ruler, as exemplified in the previous section, both for Europe and for the most advanced colony of America. The slogan of the French revolution, 'freedom, equality and brotherhood', survived the rapid monarchial reversal of power in France at the end of the 18th century and resurfaced at the 1848 revolution and subsequently became the banner inscription of the new republic. What exactly was contained in these catchwords?

Away from Paris (notable in the direction toward the Pyrenees), there was always a tendency to emphasize 'freedom', interpreted both as freedom from the centralized administration and as personal freedom. 'Equality' had primarily been associated with equal treatment before the law and in court actions, while equal access to, say, public office was considered negotiable, due to the different qualifications of people seeking such jobs. 'Brotherhood' was sometimes even omitted,[‡‡‡] but if it was there, it could mean consideration for society at large, a desire to optimize common welfare or just try to avoid conflicts between groups and fractions. The French Revolution is primarily remembered for its atrocious use of the guillotine. The 1791 'revolution' almost immediately decayed into mob rule, used by people like Danton and Robespierre to execute their personal enemies (not unlike the Greek situation after Pericles' death). Despite these circumstances of creation, the Declaration of Human Rights and the constitution based on it became the foundation for nearly all implementations of democracy during the following years and decades, and the 'Rights of Man' particularly became the point of departure for the 1948 UN declaration discussed in Section 2.3.1. The original declaration read as follows in English translation (Anonymous, 1789):

1. Men are born and remain free and equal in rights. Social distinctions may be founded only upon the general good.
2. The aim of all political association is the preservation of the natural and imprescriptible[§§§] rights of man. These rights are liberty, property, security, and resistance to oppression.

[‡‡‡]For instance in the French Declaration of the Rights of Man and of the Citizen, next page.
[§§§]That cannot be revoked, inviolable.

3. The principle of all sovereignty resides essentially in the nation. No body nor individual may exercise any authority which does not proceed directly from the nation.

4. Liberty consists in the freedom to do everything, which injures no one else; hence the exercise of the natural rights of each man has no limits except those, which assure to the other members of the society the enjoyment of the same rights. These limits can only be determined by law.

5. Law can only prohibit such actions as are hurtful to society. Nothing may be prevented which is not forbidden by law, and no one may be forced to do anything not provided for by law.

6. Law is the expression of the general will. Every citizen has a right to participate personally, or through his representative, in its foundation. It must be the same for all, whether it protects or punishes. All citizens, being equal in the eyes of the law, are equally eligible to all dignities and to all public positions and occupations, according to their abilities, and without distinction except that of their virtues and talents.

7. No person shall be accused, arrested, or imprisoned except in the cases and according to the forms prescribed by law. Any one soliciting, transmitting, executing, or causing to be executed, any arbitrary order, shall be punished. But any citizen summoned or arrested in virtue of the law shall submit without delay, as resistance constitutes an offense.

8. The law shall provide for such punishments only as are strictly and obviously necessary, and no one shall suffer punishment except it be legally inflicted in virtue of a law passed and promulgated before the commission of the offense.

9. As all persons are held innocent until they shall have been declared guilty, if arrest shall be deemed indispensable, all harshness not essential to the securing of the prisoner's person shall be severely repressed by law.

10. No one shall be disquieted on account of his opinions, including his religious views, provided their manifestation does not disturb the public order established by law.

11. The free communication of ideas and opinions is one of the most precious of the rights of man. Every citizen may, accordingly, speak, write, and print with freedom, but shall be responsible for such abuses of this freedom as shall be defined by law.

12. The security of the rights of man and of the citizen requires public military forces. These forces are, therefore, established for the good of all and not for the personal advantage of those to whom they shall be entrusted.

13. A common contribution is essential for the maintenance of the public forces and for the cost of administration. This should be equitably distributed among all the citizens in proportion to their means.

14. All the citizens have a right to decide, either personally or by their representatives, as to the necessity of the public contribution; to grant this freely; to know to what uses it is put; and to fix the proportion, the mode of assessment and of collection and the duration of the taxes.

15. Society has the right to require of every public agent an account of his administration.

16. A society in which the observance of the law is not assured, nor the separation of powers defined, has no constitution at all.

17. Since property is an inviolable and sacred right, no one shall be deprived thereof except where public necessity, legally determined, shall clearly demand it, and then only on condition that the owner shall have been previously and equitably indemnified.

Neither equality nor brotherhood is explicitly mentioned in this declaration. It is seen that the basic human rights (Article 2) are taken to include property ownership, as further expanded in Article 17. The concept of 'sovereignty' introduced in Article 3 and accorded uniquely to the state is here the monopoly on the use of violence. Later interpretations associate sovereignty with how to divide tasks between individual states and a federation or union, for example in the cases of the USA or the European Union. Article 5 limits the permitted laws to exclude arbitrary law making (a fine reminder to current politicians), and Article 6 gives the people the sole right to pass laws, whether by direct or indirect voting. Articles 7–9 provide the protection of the citizens in legal matters such as arrest and trial and Articles 10–11 ensure the freedom of speech.

Duties are introduced in the following Articles 12–14, noting the possible need for military service and taxation (or other ways in which citizens can contribute to maintaining their society). Article 13 suggests a fixed taxation rate for all. Article 15 requires accountability from civil servants and governments and Article 17 ensures that private property can be expropriated by the state if common causes are thereby furthered. The declaration is the first to make up with the earlier definitions of human rights as selfish privileges that can be enjoyed without giving anything in return. Here living in a society is seen as providing benefits conditioned on fulfilling certain obligations, without which there would be no society but just a bunch of individuals trying to maximize their own well being. Emphasizing the coupling between rights and duties is as important today as during the modern democracy formation years, especially after the appearance of the neo-liberal economic paradigm that puts personal gain above anything else.

From a contemporary point of view, there are several of the human rights and duties concepts that appear to need further qualifications. The freedom of speech article was formulated at a time with censorship on written and spoken communication. Although the concerns over such quenching of criticism are still valid, a major problem today is rather junk information and the perversity of advertising, aimed at hiding the inferiority of certain products rather than at making purchasing decisions easier, and also the kind of advertising used in political campaigns to deceive voters to put their cross in favor of politicians who may not be the least interested in the needs of the citizens, but only in securing their own livelihood. Thus, a modern formulation of the freedom of expression should subject commercial messages to strict ethical conditions, making dissemination of false or misleading information a criminal offence. One method could be the use of obligatory labelling (contents, method of production, energy consumption, environmental emissions, etc.), to be required as part of any address to the public concerning a product, along with an objective description of the usage of the product. A radical version of this could be to forbid advertising entirely and only allow a public web description of competing products, as discussed in Section 3.1.4. The same objectivity and completeness condition could be imposed on political advertisement.

The French declaration misses a precise description of equality. For example, equal opportunities are possible only if personal inheritance is abolished, and further require equal access to education and health provision, which de facto means free access, because equal opportunities do not remove inequality in society. The other missing requirement is the establishment of a brotherhood, which in practice will imply a need to emphasize the solidarity, without which the right to hold different opinions and beliefs cannot be secured. Thus, the banner tripartition might take the form *conditional liberty, increased economic equality, and first of all a deep-felt social solidarity.*

In the present context, free education should be specified as *lifelong, continued education,* because this is the requirement that our current technology and knowledge stage sets in order for it to function. The duty part of the French declaration should be generalized to require everyone to contribute to the common good according to ability and not just by paying a tax. This has been the requirement of recently proposed constitutions (Sørensen, 2012, Chapter 12) and will receive a more detailed exposition in Chapter 5.

3.2.3 Levels of Society

History has demonstrated how some smaller city states and Swiss-style local areas found it easier to work with direct democracy. Larger units such as nations have instead introduced representative democracy, which was thought to send people to parliament who still retained the knowledge of the smaller community that elected them. This has largely failed and representation has been taken over by political parties, originally based on political ideas but increasingly just being centrally managed entities with little interest beyond being in power and rewarding friends. The policies of parties is still formally based on a program with some dogma, but in practice, once elected, the party members often simply obey the wishes of certain lobby groups (often those who financed their election) and of the heads of government, influenced by similar interest groups. In election campaigns, a spiffy version of the underlying aim (the party program) may have been hinted at, or efforts may simply have been made to hide the real intentions, if the spin doctors did not think that the party program would promote election (Mair, 2013).

In recent decades, globalization has been worshipped (as requested by multinational industries) and supranational structures such as the European Union have moved decisions further away from the local communities. As a result, most Europeans do not have the surplus to even check what is going on in the distant European parliament and administration. In any case, the EU construction (originally a trade organization) neglects the most basic conditions of democracy, that laws are proposed and passed by an elected parliament before being carried out by the administration. In the European Union, the civil servants (called the Commission) propose laws, and the parliament just has to accord a lump sum annual budget for the Commission. A further Council of Ministers allows representatives from the governments of the member states to vote on the laws and regulations proposed by the Commission (EU, 2015). If the EU project has alienated the population from the political process, the US federation has done much better in making the inhabitants of each state have an interest in what goes on in the federal government. This shows that it is not the size of a supranational structure that determines its success, but the way the construction is made.

If the globalization paradigm is followed, in contrast to aiming for little-interacting regions, the problem of global equity becomes acute. The state of affairs in this respect is miserable (Ward, 1979) and more global trade can only make it worse. One issue hinted at in Section 3.1 is that environmental costs of production and overseas transportation are not considered, due to the World Trade Union's imposed 'rules of trade', but the basic problem is of course the underlying belief that one can always find or create regions with low salaries, where the goods enjoyed by the rich regions can be produced at minimal cost. This is similar to the thinking of the colonial times, that one could always find 'new worlds' with land and resources that could be exploited in favor of the rich, neglecting any indigenous population that might already inhabit such regions.

Several political observers have called for a worldwide supranational structure, a 'World Government' with more comprehensive charges than the help in avoiding warfare between individual nations that lies behind previous or existing organizations such as The League of Nations and The

United Nations. The United Nations has a structure with a large assembly where each country has one vote independent of size, plus an executive Security Council (because ensuring global peace is the primary task of the organization), where the countries with the largest weapons arsenals sit permanently and a few other countries in rotation (UN, 2015).**** All offices and organizations in the United Nations are financed by voluntary contributions from member states. Accusations of nepotism (employing the brother of some African dictator, etc.) have been made and decisions unfavorable to large countries have been avoided by threats of withdrawing funding. In any case, the United Nations has not fulfilled its objective, as many wars have erupted since World War II, and several of the Security Council nations have themselves started wars of aggression against other nations, despite not getting a UN endorsement that punishing the nation in question was justified.

Nations are a relatively new invention. They appear at different periods in time, depending on where one looks, but in Europe, they were primarily formed during the first millennium, by intricate and sometimes bloody fights typically involving a self-declared king and the local clan leaders possessing the power earlier. Apart from pleasing multinational companies, the extrapolation to make the planet one global super-state has been claimed to be a way of reducing the frequency of wars. In the current epoch, war is primarily encouraged by the arms industry, and it would seem evident that war will not go away unless arms are first removed. Arms reduction requires that security be maintained in other, nonviolent ways, which would certainly clash against paradigms of unrestricted trade of any merchandise, territorial imperialism and the pride that political leaders feel in seeing army parades and feeling superior to each other. Constitutions such as the French revolutionary one clearly identifies the need to give a monopoly of violence to the top level (in that case the state) and nobody below it, and the natural extrapolation would therefore be that only a supranational entity such as the United Nations may possess arms and use them, not the individual member nations. This idea is clearly behind the UN construction, although it is far from implemented. The Security Council can decide that a military action against a national despot bothering other nations is needed (but at present has no jurisdiction inside a nation). However, the United Nations has no standing army so certain member states must pledge to supply the forces. As mentioned the problem is that these existing national forces are also used for wars outside the consent of the United Nations, and that the arms-dealing industry is supplying arms to just about anyone willing to pay for it, from nations down to rebel groups of any kind. What

****Both the EU and the UN organization have additional institutions, for example for handling legal disputes involving the member countries.

is needed is to enforce the restriction of military power to the highest level of world government, and to forbid anyone below that level to carry arms, which of course needs an ordered transition period before it is possible. Terrorist groups and weapons-based states (including nuclear powers) must first be barred from acquiring weapons from any source, that is a comprehensive ban on weapons trade, between or inside nations, and the existing weapons must be destroyed, forcefully if needed.

Only after the problem of violence has been solved, can one hope to make the World Government a viable institution for regulating trade, environmental impacts and resource utilization, as a road to creating or maintaining human welfare across the planet.

There is an evident alternative to creating a World Government, namely to divide the world into regions that may choose different economic and social organization, use different schemes of government and accept different ambitions regarding technological development. However, this can work only if globalization is suppressed because otherwise economic predators will make use of the differences to further their own interests (use cheap labor abroad, hijack resources, export poor products to other regions, etc.). Such an Earth in separate pieces has worked for millions of years, before the exploits of colonization and international trade, and even up until today, many abuses of trade have been avoided by the system of duties and import regulations aimed at preventing destruction of indigenous industry by multinationals. Also current mass tourism is a two-sided coin: on one side, it is beneficial that people visit people in other places in order to remove prejudices and appreciate the common aspects of human culture, but on the other side, such visits have created artificial demands for things that cannot sustainably become global. Each inhabitant of India or China cannot have a $100,000 gas-guzzling car; so responsible globalization would require currently affluent countries to dispense with such displays of contempt for resources and environment.

3.2.4 Direct Versus Representative Democracy

Whether a democracy works or not depends on the insights of the voters. We have seen voting introduced in countries with poor education or countries where media manipulations exert a strong influence on voting behavior. That is not democracy. Voting makes sense only when those voting have a reasoned and comprehensive insight into the issue involved and a realistic impression of the consequences of voting in one way or another. The political setup in many countries, including both government and interests operating in the background, is such that hiding the consequences of voting in particular ways has become a national sport. Just as the ancient Greek democracy and the French revolution led to mob rule because the voters had no understanding of the larger picture, then modern representative democracies have led to oligarchic leadership, shutting out the general population from most serious decision making and at the same time creating apathy or contempt for the profession of politicians. The extra complication of political parties has removed the voters further from influence, particularly regarding what issues are being considered in their parliaments.

An early discussion of democracy based on elected representatives was made by the prolific scientist Thomas Hobbes (1588–1679). He departs from human rights such as individual freedom and natural equality but concludes that the best governance is by an absolute sovereign. Yet he sees the need for an input from civil society and concludes that the sovereign must have the consent of the people through a representative structure in what he sees as a social contract (being the first to use this concept; Hobbes, 1651). Disagreeing on the need for an autocratic monarch but supporting representative democracy, the prolific English philosopher John Stuart Mill (1806–1873) notes that many citizens in his time are fed up with the behavior of their government, and he advocates as a remedy more dialogue between citizens and their rulers as a way to approach his ideas of the greatest happiness for all (*utilitarianism*) and of individual liberty balanced by the needs of a market-based society. Yet, he was open to guild or cooperative ownership of businesses (*economic democracy*), and he thought that unlimited growth was incompatible with environmental concerns (Mill, 1861; Wikipedia, 2015d).

The discussions over the century following Mill's treatise make one thing clear: it is a fundamental condition for a representative democracy to work that it is not the same people who make the laws and administer them. A further idea voiced but rarely followed is that citizens elected to sit in a representative assembly should have fixed terms and strict rules for re-election. In ancient Greece service was for one year with the possibility of one re-election. The precise length of time is less important than to make the firm statement that being a politician is not a bread and butter profession. All experience tells that even good politicians fall into routine if not corruption, if they are allowed to regard the parliamentary job as their source of livelihood.

Some countries have the phrase in their constitutions that once elected to the parliament only the member's own conscience should guide his or her voting. This is probably fair but removes the representative function. The members of parliament no longer need to promote the matters their constituency elected them for. The situation is worse if the parliament is based on political parties, as most parties require their delegates to vote as the party leader or prime minister dictates, independent of any private 'conscience'. In fact, this is connected with the introduction of *cabinet responsibility*, meaning that a government should always be backed by a majority in parliament. This requirement was added to the constitution in several countries, after bad examples of governments ruling by 'provisional laws' not supported by parliament.[††††] In any case, all these issues point in the direction of criticizing representative democracy for having weakened or eliminated the influence of citizens on government. This, of course, becomes worse if new or altered legislation is no longer originating within the parliament, but imposed by the government or a president/ king, acting (as in the EU case mentioned earlier) both as law-giver and law-executer, in violation of the division between the law-giving, the executive and the judiciary institution or branch. This tripartite system was hinted at by Aristotle, used in the Roman Republic and later England, and named by the conservative French political philosopher Montesquieu (1748). Today it is nearly universally in use.

Why there is in the classical works on tripartite governance as well as in some current nations a king or president above the three branches is not clear. It complicates governance because there now is a novel need to divide power between the hereditary or elected top figure and the branches of legislators and of a practicing government and its civil servant helpers. In some countries, the sovereign has nearly zero power, whereas in others, he/she can influence the executive government and suggest legislation in competition with the parliament (or neglect it and rule by decree or provisional law). Perhaps the invitation to duplicate power is meant to help avoid misuse of the system for corrupt or externally lobby-imposed purposes. The sovereign is in some countries (such as the United States) head of the executive government with certain rights to veto parliamentary decisions, while in other countries, the sovereign is only signing the employment contracts for the head of government and her/his ministers/ secretaries.

The whole question of how to avoid possible misuse of positions in the tripartite structure is difficult to address and have given rise to different precautions. In addition to interpreting the laws passed by parliament, the juridical branch in some countries can question whether laws and executive decisions are in conflict with the constitution, and can handle cases of impeachment charges against elected leaders including the president (in republican states). In other national constitutions, an overseer (ombudsman) outside the government has the right to bring charges against government members for misconduct, but the power to oust the culprits remains with the parliament or the Supreme Court. The overseer is functioning for a fixed number of years and is usually a professor of law with a spotless record. Bias may be introduced by the fact that Supreme

---

[††††]The availability in many constitutions of the option for provisional laws is due to perceptions of sudden needs arising for a government to act in response to a natural disaster or a military attack by another country. Such laws were never anticipated to become semipermanent ways to counter parliamental opposition.

Court judges and overseers (where relevant) are usually appointed by the head of government that happens to be in office when a new person in the job is needed. This in itself can be misused, for example in cases where by chance a majority of judges die or retire when a particular party is in power.

It has been seen with increasing frequency that nations introduce or use existing provisions for staging public referenda in cases, where far-reaching decisions have to be made, such as yielding sovereignty to international organizations. This, among other things, makes it less likely that the decisions are reversed next time the majority in parliament changes colour. Usually, making changes in the constitution is required to involve a qualified majority (such as 2/3), being passed by two consecutive elected parliaments, or by parliament and by public referendum.

The current party or party coalition-based executive governments are as already mentioned in Section 2.3.2 at variance with several of the conditions for a representative democracy to function (Mair, 2008). They frequently introduce new legislation based on yesterday's media coverage and often on minor or specific issues that should have been left to the judiciary branch or the executive departments based on existing legislation of a more general nature. However, also the judiciary branch often has problems, usually in the form of yearlong periods for handling cases where the outcome is needed more or less instantly. A standing advisory legal body could help the administration in such cases, but the practice is rather to form committees that take years to come out with a simple report (leaving additional options for shelving cases inconvenient for a given government).

A common form of corruption in representative assemblies and party-based executive government is to allow lobbies direct access to decision makers. In certain types of governance setup it is even required that so-called stakeholders get preferred access to the legislative process. In Brussels, a whole section of town surrounding the EU Commission's quarters is devoted to industry and special interest lobbyists. But stakeholder involvement is not democracy, as noted by Blinder (1997). In recent times, the debate has continued to reveal weaknesses of the representative political system. As mentioned in Section 2.3.2, Katz (1987) identified the need to replace at least the highest-ranking civil servants when governments change. This is done in the United States, while in Europe, civil servants are in lifelong tenured positions (perhaps motivated by the need to run the country during collapse, such as in Germany at the World War II defeat). Katz and Mair supplemented each other in presenting arguments showing that party-based democracy has failed and should be abolished (Katz, 2001, 2014; Mair, 2008, 2013; Katz and Mair, 2009; Rahat et al., 2008). A similar conclusion was reached by Zakaria (2003) in his treatise on populist and delegated democracy.

Various adjustments to the way democracy is currently practiced have appeared. Bessette (1980) and Fishkin and Laslett (2003) suggest that voting should be accompanied by a deliberation phase, where issues are discussed by professionals and citizens. This appears to be little more than suggesting that voters should be knowledgeable about what they are voting on. The German sociologist Habermas (1962, 1981) criticized, like many others, the representative democracy for marginalizing input from the public sphere, through commercial and mass media manipulation of public opinion, and subsequently expressed the hope that the deliberative democracy idea can revive public participation, influenced by Max Weber's thoughts on sociology and the Bessette article. A more elaborated proposal is the 'Inclusive Democracy' idea of Fotopoulos (1997), combining direct democracy and economic democracy (leaving political decisions as well as production to the 'people', the latter as proposed by Mill) and ensuring that basic needs are satisfied for everyone (not exactly copying Russell's idea, but involving a contribution of forced work, presumably only from those capable of it, in contrast to the many other suggestions of a *basic income*, cf. Chapter 5), while leaving a free choice of doing additional work to secure nonessential consumer goods. It is acknowledged that people should somehow be given the necessary insights, including ecological ones, if the extended decision making by ordinary citizens is to work (Arapoglou et al., 2004).

New electronic options could lead to better ideas for balancing direct and representative democracy. Article 11.4 of the EU 2009 treaty (Lisbon Treaty) contains a possibility called a *citizen initiative*. It allows more than one million European citizens (with some further conditions on national distribution) to submit a proposal to the European Commission, which then, if it finds the proposal suitable, may submit it to the European Parliament and the Council. The idea was to allow Internet debate and solicitation by participants from the entire European Union. Upon implementation in 2012, a number of additional conditions were added, on the number of participants in each of seven member countries, on the kinds of suggestions that would be considered, and on a cumbersome national procedure for collecting participant endorsements, excluding Internet use across the entire European Union, and further prescribing a lengthy verification process by the Commission (Carp, 2014). Carp notes the large number of proposals actually submitted, of which none seem to have been accepted, and lists a number of initiatives refused by the Commission, demonstrating that each refusal involved a political standpoint taken by the Commission's civil servants. This initiative seems to use 'e-democracy' to please only those European citizens who have no intention to try it out. The more obvious step of allowing EU-wide referenda, by Internet or otherwise, has already been rejected at a European Convention. The Lisbon Treaty states that the European Union is based solely on representative democracy, although, as mentioned earlier, the actual EU construction is in basic conflict with most definitions of democracy.

Clearly, this negative example does not preclude that e-democracy can be used in the future. The claim is of course, that while early cases of direct democracy was limited to a city or a small number of people, then Internet usage may extend the possibilities to much larger societies, who can interact to conduct the informed debate on issues before voting upon them, even if they are physically far from each other.

In summary, neither direct nor representative democracy will function unless the citizens master sufficient knowledge to be able to understand the issues voted on and are able to ignore noise from special interests trying to influence the outcome. Furthermore, any democracy requires knowledge on the side of those ruling. If they are elected, a condition for running for office in any of the tripartite branches of government or for presidency must be both a deep knowledge of the (general or specific) subject area involved, but also a spot-free record of incorruptible personal behavior and soberness in any previous political activities.

References

Andersen, A., Christensen, A., Nielsen, N., Koob, S., Oksbjerg, M., Kaarup, R., 2012. The wealth and debt of Danish families. Monet. Rev. (of Danish National Bank) 2, 1–40, 2nd Quarter.

Anonymous, 1215. Magna Carta. Available from: www.gutenberg.org (Ebook # 10000).

Anonymous, 1789. Déclaration des droits de l'homme et du citoyen, French National Constituent Assembly. English translation, Declaration of the rights of man and of the citizens. Available from: www.hrcr.org/docs/frenchdec.html

Anthony, P., 1977. The Ideology of Work. Tavistock Publisher, London.

Arapoglou, P., Fotopoulos, T., Koumentakis, P., Panagos, N., Sargis, J., 2004. What is inclusive democracy? Opening statement of editors. Int. J. Inclus. Democr. 1 (1), 1–9, www.inclusivedemocracy.org/journal/

Aristotle, 326 BC. Politika. Particularly the possibly unfinished Books 7–8. English translation, Politics by W. Ellis, 1912. This is believed to be part of extended lecture notes on several different subjects, written by Aristotle between 335 and 323 BC. Available from: www.gutenberg.org (Ebook # 6762).

Aristotle, 328 BC. Athenaion Politeia. May have been written by a student of Aristotle as research for the book Politika. English translation, The Athenian Constitution by F. Kenyon. Available from: www.gutenberg.org (Ebook # 26095).

Arrow, K., Dasgupta, P., Goulder, L., Mumford, K., Oleson, K., 2012. Sustainability and the measurement of wealth. Environ. Dev. Econ. 17, 317–353.

Bakunin, M., 1882. God and the State. Fragment. First US edition by Mother Earth Publ. Assoc., New York, 1916. Available from: www.marxists.org/reference/archive/bakunin/works/godstate/index.htm

Bessette, J., 1980. Deliberative democracy: the majority principle in republican government. How Democratic is the ConstitutionAEI Press, Washington DC, 102–116.

Blinder, A., 1997. Is government too political? Foreign Aff. 76 (6), 115–126.

Bolgar, R., 1958. The Classical Heritage and its Beneficiaries. Cambridge University Press, Cambridge, (updated version 1973).

Carp, R., 2014. L'Initiative citoyenne Europeenne et les vertus de la democratie directe sous la forme de l'e-democratie. Studia Europaea (Universitatis Babes-Bolyai, Romania) 59, 147–171.

Ceriani, L., Verme, P., 2012. The origins of the Gini index: extracts from Variabilità e Mutabilità (1912) by Corrado Gini. J. Econ. Inequal. 10, 421–443.

Cicero, M., 54 BC. De re publica. Only a fragment is preserved. English translation, On the commonwealth. Available from: www.gutenberg.org (as part 3 of Cicero's Tusculan Disputations, etc., Ebook # 14998).

Clarke, W., Woodhouse, A., 1938. Clarke originally recorded the 17th century discussions, Woodhouse edited the 1938 edition. Puritanism and Liberty, being the Army Debates (1647–1649) from the Clarke Manuscripts with Supplementary Documents. Available from: oll.libertyfund.org/title/2183

Danish Ministry of Food, Agriculture and Fisheries, 2010. Statistics on organic farms 2009. Plantedirektoratet, Copenhagen, Available from: pdir.fvm.dk

Danish National Bank, 2015. Statistical databases DNFK, DNSOSB, DNMNOGK. Available from: nationalbanken.statistikbank.dk.

Davies, J., Sandström, S., Shorrocks, A., Wolff, E., 2009. The level and distribution of global household wealth. US National Bureau of Economic Research working paper.

Downing, B., 1989. Medieval Origins of Constitutional Government in the West. Theory Soc. 18 (2), 213–247.

EC, 1995. ExternE: Externalities of Energy. 5-Volume project report from DG XII, Luxembourg, ISBN 92-827-5212-7.

EC, 2013. Progress on "GDP and Beyond" actions. Staff working document, European Commission, Brussels.

EC, IMF, OECD, UN, WB, 1993. System of National Accounts 1993. Commission of the European Communities, Brussels.

Elazar, D., 1993. Communal democracy and liberal democracy: an outside friend's look at the Swiss political tradition. Publius 23 (2), 3–18.

EU, 2015. How the EU works. Available from: europa.eu/about-eu/

Fishkin, J., Laslett, P., 2003. Debating Deliberative Democracy. Wiley, Blackwell, London.

Fotopoulos, T., 1997. Towards an Inclusive Democracy: The Crisis of the Growth Economy and the Need for a New Liberatory Project. Cassell, London.

Galbraith, J., 1976. The Affluent Society, third ed. Houghton Mifflin Co., New York.

Gemynthe, F., 1998. Rigdom Uden Arbejde. Chr. Ejler's Forlag, Copenhagen, (in Danish).

Gini, C., 1912. Variabilità e Mutabilità: Contributo Allo Studio Delle Distribuzioni e Delle Relazioni Statistiche. Tipografia P. Cuppini, Bologna.

Gravgård, O., 2013. Grønne nationalregnskaber og det grønne BNP. Statistics Denmark, Copenhagen (in Danish).

Habermas, J., 1962. Strukturwandel der Öffenttlichkeit. English translation, The Structural Transformation of the Public Sphere. Polity, Cambridge.

Habermas, J., 1981. Theorie des Kommunikativen Handelns. English translation, The Theory of Communicative Action. Beacon Press, Boston.

Hansen, M., 1989. Was Athens a Democracy? In: Historisk-Filosofiske Meddelelser, vol. 59. Royal Danish Academy of Sciences and Letters, Copenhagen.

Hansen, M., 1999. The Athenian Democracy in the Age of Demosthenes. Structure, Principles and Ideology, second ed. Bristol Classical Press, London.

Hansen, M., 2013a. Plato. Jurist og Økonomforbundets Forlag, Copenhagen.

Hansen, M., 2013b. Reflections on Aristotle's Politics. Museum Tusculanum Publ., Copenhagen.

Hartwick, J., 1990. Natural resources, national accounting and economic depreciation. J. Public Econ. 43, 291–304.

Hobbes, T., 1651. Leviathan. Available from: www.gutenberg.org (Ebook # 3207).

Illich, I., 1978. The Right to Useful Unemployment. Marion Boyars, Boston.

Katz, R., 1987. Party Governments: European and American Experiences. De Gruyter, Berlin.

Katz, R., 2001. Models of democracy. Eur. Union Polit. 2 (1), 53–79.

Katz, R., 2014. No man can serve two masters: party politicians, party members, citizens and principle-agent models of democracy. Party Polit. 20 (2), 183–193.

Katz, R., Mair, P., 2009. The cartel party thesis: a restatement. Perspect. Polit. 7 (4), 753–766.

Larsen, J., 1949. The origin and significance of the counting of votes. Classic. Philol. 44 (3), 164–181, Available from: www.jstor.org.

Lee, R., 1979. The !Kung San: Men, Women, and Work in a Foraging Society. Cambridge University Press, Cambridge.

Lewis, J., 2011. Constitution and fundamental law: the lesson of classical Athens. Soc. Philos. Policy 28 (1), 25–49.

Locke, J., 1689. Two treatises of government. Available from: oll.libertyfund.org /titles/222

Lutz, D., 1998. Colonial Origins of the American Constitution. Liberty Fund, Indianopolis. Available from: http://lf-oll.s3.amazonaws.com/titles/694/Lutz_0013_EBk_v6.0.pdf

Lyttkens, C., 1994. A predatory democracy? an essay on taxation in classical Athens. Explor. Econ. Hist. 31, 62–90.

Lyttkens, C., 2010. Institutions, taxation, and market relationships in ancient Athens. J. Inst. Econ. 6 (4), 505–527.

Mair, P., 2008. The challenge to party government. West Eur. Polit. 31 (1–2), 211–234.

Mair, P., 2013. Ruling the Void – The Hollowing of Western Democracy. Verso, London.

Markoff, J., 1999. Where and when was democracy invented? Comp. Stud. Soc. Hist. 41 (4), 660–690.

Marx, K., Engels, F., 1848. Manifesto of the Communist Party. Educational Society, London (originally in German). Annotated version on Marxists.org

Mill, J., 1861. Considerations on representative government. Available from: www.gutenberg.org (Ebook # 5669).

Montesquieu, 1748. Baron C-L de Secondat. De l'Esprit des Loix (The Spirit of the Laws). Barillot & Fils, Geneve. Available from: www.gutenberg.org (Ebook # 27573).

Nordhaus, W., Tobin, J., 1972. Is growth obsolete? In: Economic Research: Retrospect and Prospect, vol. 5. Economic Growth. US National Bureau of Economic Research, p. 80. Available from: www.nber.org/chapters/c7620

OECD, 2014. Green Growth Indicators 2014. Organisation for Economic Co-Operation and Development, Paris.

Ohlsson, H., Roine, J., Waldenström, D., 2006. Long-run changes in the concentration of wealth. UNU-WIDER Research Paper 2006/103. United Nations University, pp. 1–32.

Orth, S., 1913. Socialism and Democracy in Europe. Henry Holt & Co., New York, Available from: www.gutenberg.org (Ebook # 35572).

Paine, T., 1776. Common Sense. Available from: www.gutenberg.org (Ebook # 147).

Piketty, T., 2014. Capital in the Twenty-First Century. Harvard University Press, Boston.

Piketty, T., Saez, E., 2013. Top incomes and the great recession: recent evolutions and policy implications. IMF Econ. Rev. 61 (3), 456–478.

Piketty, T., Saez, E., 2014. Inequality in the long run. Science 344 (6186), 838–843.

Piketty, T., Postel-Vinay, G., Rosenthal, J.-L., 2006. Wealth concentration in a developing economy: Paris and France 1807–1994. Am. Econ. Rev. 96 (1), 236–256.

Plato, 345 BC. Laws. (B. Jowett, Trans.) (in English). Available from: www.gutenberg.org as Ebook # 1750. This book is from the late part of Plato's life, because it is regarded as unfinished.

Plato, 380 BC. The Republic. (B. Jowett, Trans.) (in English). Available from: www.gutenberg.org as Ebook # 1497. This book is from the middle period of Plato's working life, and by its contents after the Athenian execution events of 406 and 399 BC.

Rahat, G., Hazan, R., Katz, R., 2008. Democracy and political parties. Party Polit. 14 (6), 663–683.

Rousseau, J-J., 1755. Discours sur l'origine et les fondements de l'inégalité parmi les hommes. Available from: www.rousseauonline.ch; English translation, A Discourse Upon the Origin and the Foundation of the Inequality Among Mankind. Available from: www.gutenberg.org (Ebook # 11136).

Rousseau, J-J., 1762. Du contrat social, ou principes du droit politique. Available from: www.rousseauonline.ch; English translation, The Social Contract. Available from: www.gutenberg.org (Ebook # 46333).

Rubin, E., 2001. Getting past democracy. Univ. Penn. Law Rev. 149 (3), 711–792.

Schwartzkopff, F., 2013. Danes as most-indebted in world resist credit: mortgages. Available from: www.bloomberg.com/news/articles/2013-04-28/danes-as-most-indebted-in-world-resist-credit-mortgages

Skat, 2015. Ejendoms- og grundværdier 2001–2012 for alle typer ejendomme: landsniveau. Danish Tax Authority. Available from: www.skat.dk.

Solow, R., 1986. On the intergenerational allocation of natural resources. Scand. J. Econ. 88 (1), 141–149.

Sørensen, B., 2010. Renewable Energy, fourth ed. Academic Press/Elsevier, Burlington, MA.

Sørensen, B., 2011. Life-cycle Analysis of Energy Systems – From Methodology to Applications. RSC Publishing, Cambridge.

Sørensen, B., 2012. A History of Energy – Northern Europe from the Stone Age to the Present Day. Earthscan/Routledge, Cambridge.

Sørensen, B., 2014. Energy Intermittency. CRC Press, Taylor & Francis, Boca Raton.

Statistics Denmark, 1919. Ansættelserne til Indkomst- og Formueskatten for Skatteåret 1918/19. Statistiske Meddelelser fourth series, vol. 57, booklet 7. Department of Statistics, Bianco Luno Publ., Copenhagen.

Statistics Denmark, 1938. Indkomst- og Formueansættelserne til Staten for Skatteåret 1937/38. Statistiske Meddelelser fourth series, vol. 106, booklet 6. Department of Statistics, Bianco Luno Publ., Copenhagen.

Statistics Denmark, 1975. Indkomster og Formuer ved Slutligningen for 1972. Statistisk Tabelværk, IV. Copenhagen.

Statistics Denmark, 1998. Income and Wealth 1996. Copenhagen.

Statistics Denmark, 2007. Indkomster 2005. With Partial Wealth Information. Copenhagen. Pdf version available from: www.dst.dk

Statistics Denmark, 2014. Indkomster 2012. Copenhagen. Pdf version available from: www.dst.dk

Statistics Denmark, 2015a. Input–Output Tables 1966–2011. Available from: www.dst.dk/en/Statistik/emner/produktivitet-og-input-output/input-output-tabeller.aspx?tab=tab#

Statistics Denmark, 2015b. Geography, Environment and Energy. Energy Accounting. Statistical databases ENE2HT, ENE2HA, MRM2. Available from: www.statistikbanken.dk; www.dst.dk

Statistics Denmark, 2015c. National Accounts and Government Finances. Statistical databases NASK, NAHF, UDG12, REGK63. Available from: www.statistikbanken.dk

Statistics Denmark, 2015d. Money and Credit Market. Statistical database MPK49. Available from: www.statistikbanken.dk

Statistics Denmark, 2015e. Education and Knowledge. Statistical database KRHFU2. Available from: www.statistikbanken.dk

Statistics Denmark, 2015f. Labour, Earnings and Income. Statistical database INDKP7/31/107. Available from: www.statistikbanken.dk

Statistics Denmark, 2015g. Populations and Elections. Statistical database HISB7. Available from: www.statistikbanken.dk

Stieglitz, J., Sen, A., Fitoussi, J.P., 2009. Report by the Commission on the Measurement of Economic Performance and Social Progress. Report to the French President. Available from: http://www.stiglitz-sen-fitoussi.fr/documents/rapport_anglais.pdf

Thoreau, H., 1754. Walden, and on the duty of civil disobedience. Available from: www.gutenberg.org (Ebook # 205).

Thucydides, 431 BC. The History of the Peloponnesian War (R. Crawley, Trans.) (in English). Book II, Chapter 6, Part 3. It is not known how verbatim the rendering of Pericles' speech is. Available from: www.gutenberg.org Kindle Locations 1702–1711.

Tocqueville, A., 1835, 1840. Democracy in America, vols. I and II. Available from: www.gutenberg.org (Ebooks 815 and 816).

UN, 2006–2014. Millennium Development Goals. Progress reports on meeting the goals of the UN Millennium Declaration. UN Department of Economic and Social Affairs, Statistics Division. Available from: mdgs.un.org/unsd/mdg/Default.aspx

UN, 2015. United Nations main organs. Available from: www.un.org

UN, EU, FAO, IMF, OECD, WB, 2014. System of Environmental-Economic Accounting 2012. Central Framework. United Nations, New York.

US EPA, 2015. Frequently asked questions on mortality risk valuation. National Center for Environmental Economics, US Environmental Protection Agency. Available from: yosemite.epa.gov/ee/epa/eed.nsf/webpages/mortalityriskvaluation.html

Viscusi, W., Aldi, J., 2003. The value of a statistical life: a critical review of market estimates throughout the world. J. Risk Uncertain. 27 (1), 5–76.

Voltaire, 1759. Pseudonym for F.-M. Arouet. Candide. Available from: www.gutenberg.org (Ebook # 147).

Ward, B., 1979. Progress for a Small Planet. Norton & Co., New York.

Weltzman, M., 1976. The optimal development of resource pools. J. Econ. Theory 12, 351–364.

Wiessner, P., 2002. Hunting, healing, and *hxaro* exchange. A long-term perspective on !Kung (Ju/'hoansi) large-game hunting. Evol. Hum. Behav. 23, 407–436.

Wikipedia, 2015a. Mikhail Bakunin. Available from: en.wikipedia.org

Wikipedia, 2015b. Jean Jaurès. Available from: en.wikipedia.org

Wikipedia, 2015c. Legislative assemblies of the Roman Republic. Available from: en.wikipedia.org

Wikipedia, 2015d. John Stuart Mill. Available from: en.wikipedia.org

World Bank, 2006. Where is the Wealth of Nations? – Measuring Capital for the 21st Century. Washington, DC.

World Bank, 2011. The Changing Wealth of Nations – Measuring Sustainable Development in the New Millennium. Washington, DC. Tabular data available in spreadsheet format from: data.worldbank.org/data-catalog/wealth-of-nations

World Resources Institute, 2003. Millennium Ecosystem Assessment: Ecosystems and Human Well-being: A Framework for Assessment. Island Press, Washington, DC. Available from: pdf.wri.org/ecosystems_human_wellbeing.pdf

Zakaria, F., 2003. The Future of Freedom: Illiberal Democracy at Home and Abroad. Norton & Co., New York.

Zimmerman, J., 1999. New England Town Meeting: Democracy in Action. Greenwood Press, Boston.

Specific Issues and Case Narratives

4.1 ENERGY EFFICIENCY AND INFRASTRUCTURE

The need to make the activities taking place in societies sustainable entails particular requirements for the systems providing and using energy, because energy is a common denominator for nearly all types of manufacture, service and many types of welfare activities. As energy systems are also among the best-studied subsystems in our societies, it is natural to use them as examples for looking at the difference between our available insights and our current behavior and practices. In this section this will be done for the efficiency of energy conversion, a property relevant for any type of energy use (a sloppy word for conversion from a higher to a lower quality energy form). Efficiency may be increased either by optimizing the currently employed energy conversion processes or by introduction of new, better processes for reaching the same ends.

Once the energy system is made efficient an appropriate mix of energy sources that will fulfil our social and environmental requirements has to be found. Generally the cost of introducing any kind of sustainable new energy source is considerably higher than investing in an equivalent efficiency improvement. This means that we are so far from using energy efficiently today that measures of small cost can improve the situation. A four to six times efficiency improvement can be obtained by measures that are cheaper than the cost of the energy they save (Weizsäcker et al., 2009; Sørensen, 2010, 2012a).

After the investment in energy efficiency has been made, there will be more options for satisfying the now lower energy input requirements. However, seen from an environmental and human health perspective, some avenues have to be counted out. Fossil fuels create pollution which can often be managed, but the emissions of greenhouse gases cannot, so

Energy, Resources and Welfare. http://dx.doi.org/10.1016/B978-0-12-803218-3.00004-5

fossil energy should be left for noncombustion uses (plastic, lubricants, etc.), even if there are resources left. Ideas of decarbonizing fossil fuels such as coal have little merit because it means transforming coal to something else, which is a complex process. The energy in the coal has to be transferred to an acceptable fuel, such as hydrogen, but the enormous amounts of waste products from the process (presumably in the form of CO_2) must be handled, for example by storage at the bottom of the ocean. It is easy to estimate that the cost of this scheme is higher than producing the same amount of hydrogen by electrolysis from renewable energy such as photovoltaic or wind power, the flow of which cannot always be used directly when produced (Sørensen, 2012a, 2014). Conventional arguments that hydrogen is cheaper to produce from natural gas than from wind are of course wrong because they omit the cost of the global warming impacts.

Nuclear power also fails the acid test. Everyone in modern society knows and accepts that it is human nature to make mistakes. Therefore, they must reject use of technologies that require 100% error-free operation. A contemporary person will probably also reject the historical Japanese virtue of never losing face as, for example when the people in the electric utility company running the Fukushima reactor chose to hide their inability to correctly translate the US reactor manufacturer's safety manual by simply omitting certain sections, causing the erroneous reaction of the operators in the control room that triggered the 2011 disaster (Elliott, 2013; Sørensen, 2014).

From a life-cycle perspective, the least expensive energy systems are those based on wind, solar or biomass sources. The latter have environmental impacts of concern, while solar and wind technologies have only negligible impacts. System options such as using hydrogen to store surplus renewable energy to use in periods with insufficient production have been devised (Sørensen, 2012a, 2014). The following sections will discuss the energy efficiency issues from a perspective of the policy measures needed for promoting the economically viable options.*

4.1.1 Buildings and Commerce

The energy use of buildings is a substantial part of the energy picture in many countries. In some countries, substantial amounts of space heating is required during winter, and in (usually) other countries, space cooling is needed during summer. To this comes the energy use within the building, comprising hot water, power for light, computers and other equipment. In

*The strange fact that many customers prefer to invest in new energy supply rather than the lower-price efficiency measures may have a psychological explanation in that it is felt as more flashy to exhibit a new solar panel on the roof than buying a car with a more efficient engine, especially in a world where visible 'growth' is constantly advertised.

both dwellings and commercial buildings there would further be energy consumption by refrigerators and deep freeze compartments, in offices various other technical equipment and in buildings used for manufacture additional machinery and heat-based processing in boilers and furnaces. At present, the selection of equipment for these energy-converting types of equipment is basically done on the basis of some cost estimate. As noted in Section 3.1.4 this has in some cases led to selection of the most efficient equipment, but not in other cases.

For the buildings themselves, establishing efficient energy behavior involves passive factors of the architecture, the materials used for the building shell as well as the organization of activities within the building. Architects are supposed, in contrast to artists, to have the comfort of the inhabitants as their primary concern. Unfortunately, this is often not the case. The architect designs a structure he feels is pleasing, and only then calls in the energy-knowledgeable engineer to make the structure realizable, thereby excluding the optimization opportunity that incorporating energy consideration already in the early design phase could have provided. Extreme examples of this kind of failure are seen in buildings with glass facades placed in climates with long periods of high temperature and with high incident solar radiation. Attempts to reduce the heating of the inner rooms from absorption of radiation through the windows may be made through smart window technologies (changing reflectivity with radiation input) and with Venetian blinds, but also when such features are used, the indoor temperature often becomes uncomfortable and active cooling is thus installed, and often in excessive amounts, causing visitors to the building to catch colds when entering from the hot outside to the 'frosty' inside. The fact is that proper passive designs can eliminate the need for active cooling in nearly all climates and for nearly all periods of the year. A few hours a year may require cooling, but far from the dominating contribution to energy usage seen at present. Methods include the right choice of materials, providing light by more sophisticated routes than Sun-facing windows, proper design of exterior walls and inside rooms, as well as active cooling methods by intelligent use of water and moisture flows, solar and wind-driven passive or active devices, for example solar chimneys in dry climates and dehumidifiers in wet climates (Bahadori, 1978; also as Chapter 9 in Sørensen, 2011a, where other articles on efficiency are in Chapters 26–36).

The basic measure needed to keep a building comfortable is to insulate the shell. This will reduce heat losses during cold periods, but if used correctly it also will help keeping the inside of the house cooler than the outside in summer. The basic idea is to introduce an intermediate layer in each wall, consisting of a material with many tiny cavities filled naturally by air (that has a low heat coefficient) or filled with a specific substance with insulation properties better than air, but in that case having to create a long-lasting barrier to prevent the gas from escaping. Because a wall should also control humidity flows, gas filling may be inconvenient and currently is only used for transparent parts of the shell, such as windows. The philosophy presently prevailing is to make the building shell very tight although not impenetrable. This means that air circulation (replacing air containing health damaging particles from human skin, cooking and other sources by outdoor air) is essential and thus a ventilation and heat exchange system (to recuperate heat from reject air) is usually part of the building energy system. The previously used acceptance of holes in the shell relied on wind to exchange air, and therefore did not work when wind speeds happened to be low.

Many materials may be used for building walls from metal and concrete to bricks and wood. For smaller buildings, the best solution is to use a sandwich structure where the main part of the wall is insulation material, but with an outside and inside surface of either wood or one of the solid mineral kinds, weatherproof if used on the outside but usually not for the inside wall, that can be made, for example of gypsum. The insulation material between the surfaces can be stone-, glass- or paper wool, or certain pearl-bed plastic materials. Contemporary versions of these employ solutions where structural strength is preserved with inside beams filling only part of the volume, in order to avoid the 'cool bridges' where heat could be lost due to the higher transmission in the structural material compared with the insulating material. Proper design considers any heat transfer, whether though the building surface or by air exchange. In order to construct a building today with optimal energy use over its often very long lifetime (with partly unpredictable energy prices), it may well be needed to include a 0.5-m thick layer of insulation, which is easily made part of the design for a new building, but poses challenges for retrofitting older, existing buildings. Basically, adding more insulation material to an existing building wall can be made from the inside or from the outside. Inside insulation is easiest but reduces the indoor area, while outside addition presents architectural problems in creating a new façade of appealing character.

The question is how far labelling and voluntary actions can go in reaching the state of affairs desired from a societal point of view. There will always be citizens reluctant to follow any advice, and as regards existing buildings, some efficiency improvements were done after the 1973/74 energy crisis, but presently efficiency seems to be low on the list of accepted expenses, most of which go in the direction of improving the visual appearance of the building. The simplest way of ensuring that efficiency is improved is to implement regulation, similar to the one existing in most countries aimed at ensuring structural safety of the building. I shall use the Danish regulation as an example of incorporating energy efficiency in the building codes and present suggestions for a gradual improvement of the standard of the already existing building stock. The weakness of the Danish building codes is that they only apply to the construction of new buildings (and recently to major renovation or additions), and because average building lifetimes are exceedingly high in Denmark (over 100 years), the focus on new buildings means that it will take very long to change the overall energy efficiency of the entire building stock. For this reason, the discussion must be extended to discuss measures for existing buildings.

Consider a one-family detached house. Its utility may be assumed proportional to the total floor area, in the present Denmark typically in the range of 100–200 m^2. The optimal shape of the house is of course spherical, from a surface area to volume point of view, if the lowest heat loss for a given type of surface is the goal. In practice, the situation is more complex. From an indoor architectural point of view, the spherical shape may not be attractive. Furthermore, heat loss is not uniform, but may be different for surfaces facing the ground and air, to the sides or upward, depending on soil type, wind profiles and shadowing objects such as trees or other buildings. The circular profile may for reasons of indoor room design be replaced by a rectangular shape (without extrusion and irregularities contributing to heat loss), but for the upper-end floor area requirement, a two-storey building would be clearly advantageous (closer to spherical shape). Replacing spherical surfaces by flat ones also entails the advantage of being able to install solar panels (other than the flexible surface type) on façades or roof slopes. The best solution will be one that has surfaces suited to solar panel integration facing the appropriate direction (south on the northern hemisphere, north on the southern). The design of rooms for various purposes and placement of windows may also make use of such orientation. Rooms intended for sleep should be placed away from the main solar incidence direction and have few windows, rooms intended for daytime use on the opposite side. This would be true for the case of Denmark but other arrangements would be preferable closer to the Equator, where cooler rooms may also be preferred during daytime. Terrace space for outdoor eating and socializing could be on the rooftop or extending from a side of the house, or it could be courtyard-style spaces in the middle of the building, in climates where such space needs to be cool.

The Danish building code is altered about every 5 years, making new requirements known to the house constructors in advance, so that they can adapt their designs, materials and processes smoothly. The allowable energy use has diminished with time, as detailed in the caption to Fig. 4.1. Over the years, more details of building energy use and energy loss have been considered, as their importance has become apparent. For instance, the 1973/74 energy supply crisis made people perform several energy efficiency measures that could be done easily, such as attic insulation and closing crevices around windows. Subsequently it was realised that some

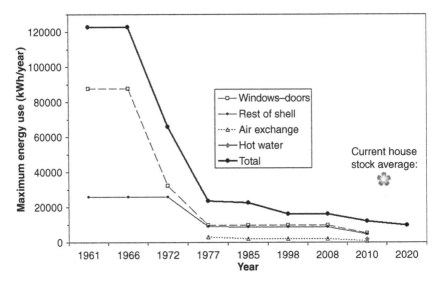

FIGURE 4.1 Danish building code prescriptions for a new 200-m² floor area, two-storey detached house with 30 degree roof inclination and 50 m² doors and windows (shell area 500 m²), as modified over time. The 1961 building code only considered heat loss through the building shell, but from 1972, also the air exchange was regulated. The 1998 and 2008 code worked with a total energy allowance per unit of floor area, but now for heating, cooling, air exchange and hot water supply together, while from 2010, a range of different rules were brought into play, the overall allowance maintained, but with specific limits for the heat loss through the building shell excluding doors and windows, and specifications for ventilation with required heat recovery and for special heat losses where different building elements come together. Based on Danish Energy Agency (2015), with links to previous building codes. The flower point indicates the average energy use of the present house stock (but for comparison adjusted to 200-m² floor size, which is higher than the actual average).

citizens overdid this so that they did not get the minimum indoor air exchange needed, in particular, for concrete-element buildings with radon emissions, but also in other houses due to bedroom dust and the like. Following this experience, the building code requirement for air exchange in kitchens and wet rooms were in 1977 set at 35 L/s, reduced to 25 L/s in 1985 and in 2010 supplemented by a requirement for cross-in/out heat exchange of an efficiency over 70%, implying heat uses of roughly 3000, 2150 and 645 kWh/year for the 200 m² building considered in Fig. 4.1.

A particular problem has been heat loss through windows, much larger and displaced in time compared with solar input through the windows, and typically the heat loss through a window is 10 times higher than through a well-insulated wall per unit area. Traditional Danish buildings have had window areas around 10% of the floor area, but after Danes got into mass tourism to the Mediterranean region, architecture suited for that region became imported to Denmark, where people hoped to get a bit more indoor light during the dark 56°N winters. In reality, significant increase in light transmission to the interior happens only in summer, where it is not needed but rather induces a perceived need for space cooling. Although window technology still causes losses much higher than through walls, the permitted window area in the building codes has now increased from 15% of the floor area in 1977 to 30% in 2010. Despite flaws such as these, the impact of the regulation on new buildings has been generally positive, as exemplified by the significant reduction in heat losses seen for the 200 m² detached house (with doors and windows amounting to 10% of the building shell area, 25% of the floor area) illustrated in Fig. 4.1. In addition to the required minimum energy usage, the present building code recommends a 'low-energy alternative', which may point to future reductions, particularly for the insulated walls.

A building energy labelling system required existing buildings to be rated and labelled A–F, and each building had to be assessed before a sale, with a report stating the rating, compulsory to be shown in any real estate advertisement for the house. The disappointing outcome has been that sales prices are practically the same for similar houses with similar location but independent of the energy rating. The standing stock of Danish buildings have a median rating between D and E, corresponding to an average energy loss through the building shell of around 190 kWh/year per m² of floor space, which is more than three times higher than the allowed maximum for new houses. As some building companies have pointed out, this means that it would be economically advantageous in a life-cycle perspective to tear down half the buildings in Denmark and replace them by new ones. This argument does not seem to include the energy content of building materials and energy usage by equipment during demolition and new construction, but even if these were considered, a very large portion of Danish houses should probably be condemned rather than polished up (visually) and resold.

Cooking ranges and appliances for cooling and freezing, clothes and dish washing and drying have long had a labelling system, and as mentioned in Section 3.1.4 this has worked well in squeezing low-efficiency products out. Also for computers and related equipment, the energy efficiency has been systematically rising, but in this case not by regulation but simply because the miniaturization trend has forced the manufacturers to use the solutions with least waste heat as such heat will damage components and in some cases shuts the computer down (a safety feature usually set to go off just below 100°C). This does not mean that more cannot be done and should be done. For example the trend toward very large screens for television sets and stationary computers has almost nullified the reduction in energy consumption happening when cathode ray screen technology was replaced by solid-state solutions. At the other end, for small, portable computers and phones, the stuffing of the apparatus with features has made the interval needed between battery recharging decline from nearly a week to often under a day. A major culprit is again the larger displays. In the case of mechanical devices from lawn movers to saws and cutters, a transition has been taking place from fuel dependency to battery operation, here with fairly acceptable recharging intervals because such equipment is only used for relatively short durations. The other main energy user in homes, offices and shops is light. Here the transition from incandescent and halogen light bulbs to compact fluorescent lamps and light-emitting diode lights has greatly reduced energy use (by factors of 4–7).

In retail commerce, a large amount of energy is used for cool and frost compartments, because these are open for the shoppers to see the goods and therefore the efficiency of well-insulated, closed containers is not reached. Some improvement is made by providing glass covers that the customers have to push open in order to access the merchandise. However, as in the case of windows, this does not help much due to the poor insulation characteristics of glasses, as even double layered glass transmit large amounts of heat (and these are not used in shops, being satisfied with just preventing some of the cold air from escaping).

4.1.2 Transportation and Services

The transportation sector has, for several decades, been sporting one of the highest growth rates in number of vehicles and kilometres driven. In some countries, this has recently changed into a more differentiated picture, where purchases of large vehicles for transportation of goods have moved toward the most energy efficient lorries, while purchases of small trucks and vans for the service sector have gone in the opposite direction. For passenger vehicles, sales are split between a group of customers buying the most efficient cars on the market and another group that demonstratively buys the most extravagant cars they can find. This is due to the fact that in certain countries cars are still regarded as a status symbol rather than a means of transportation and as mentioned in Section 3.1.4, in some cases customers for these vehicles wants to display their disregard for the environment and show that they can afford to drive a gas-guzzler. Still, some manufacturers of limousines, sports cars and special utility vehicles have started to also improve their energy efficiency, showing more social respect than their customers.

Again, Denmark may be taken as an interesting case story, because of its substantial car taxation. In part, there is a progressive registration tax of 105% of the bare price for inexpensive cars, rising to 180% for the most expensive ones, with an efficiency-dependent correction. In addition, an annual tax is levied on all cars but graduated according to their energy efficiency, based on the certified European Union test drives along a prescribed mixed urban/nonurban driving cycle. For the most efficient cars, this tax is about €70, and rises with falling energy efficiency to over €1000 for the worst performers. Sales outlets are obliged to mention EU rating and tax in all advertisements. The significant sales of luxury cars support the remarks on customer behavior made earlier. It has been suggested that taxation should be based on a more precise parameter than the average kilometres per litre of fuel, causing confusion when comparing gasoline, diesel, electric and hybrid cars, or cars with different payloads. Sørensen (2007) suggested that instead, taxation or regulation should be based on energy used per unit of mass moved a kilometre and this would solve the problems mentioned earlier. The product of payload (p) and distance (d) is the *transport work* performed, so if the energy used is e, then the vehicle's efficiency may be expressed by a performance index, i, written as

$$i(\text{kg km} / \text{MJ}) = p\,(\text{kg}) \times d\,(\text{km}) / e\,(\text{MJ}).$$

This definition can be used for passenger vehicles of any type and size (from one person and up plus allowed filling of luggage compartments) as well as for freight transportation vehicles (now including the cargo mass in the payload) and any other vehicle (trains, airplanes, ships). In consequence, it allows all regulation to be performed with a single requirement, such as allowing only vehicles with $i > 350$ km kg/MJ. This condition is fulfilled for several available passenger cars, heavy trucks, freight trains and cargo ships, but not for current airplanes.

In many countries, there is competition between public and private transportation, where public transportation is defined as including passenger transport by rail, sea, air or road, comprising both buses and taxis. The railed systems had big success in previous centuries, being faster and likely cheaper than competition from stagecoaches, horse carriages or riding. Buses only could make an indent on the trains after suitable road networks have been built, by offering a denser coverage of destinations. However, as private cars have reached high ownership fractions, the convenience of door-to-door transport has made them take over the bulk of passenger transportation, leaving occupancy percentages in trains and buses at such low values that schedule frequencies have had to be diminished (making these means of transportation less attractive) and prices have gone up, causing further reduction in the number of users (although, strictly calculating, most public transport is in fact still cheaper than driving a private car, at least if there is only one person in the private car).

There are very good arguments to suggest that public transportation should be free. The cost of building and maintaining the road system with dense road networks in populated areas plus regional highways and motorways, including traffic signs, land use, environmental disturbances, noise and visual impacts, in many countries would exceed the revenues from car taxation, and thus it is quite unfair if taxpayers in this way subsidise the use of private cars but not public means of transportation. In any case, having paid (often long ago) the expense of establishing railway tracks it seems rather silly to charge tickets for using them when that scares the customers away.

Bicycles have experienced periods of very high usage, for commuting, shopping and other limited distance trips. This stopped in some countries when people started to settle far from their workplaces and when community shops were replaced by supermarkets, hypermarkets and shopping centres sited on cheap land. Today, most cities are unsuited for motorcars due to pollution and congestion, and in Europe many city centres has been declared off-limits for cars. Therefore, bicycles are again on the rise, both private bicycles for local people and 'white' public bicycles for free usage by tourists and other town visitors (in some cities users are unfairly being charged money for using the cycles, despite the general interest of cities to attract visitors). Cars are also most energy inefficient for short 'shopping' trips, so insisting that bicycles should be used by all people fit to ride, for shorter trips, is one important step toward making transportation sustainable. It should also be considered to curb the placement of shopping centres and hypermarkets so far away from population accumulation areas that many will deselect bicycles.

There is another relationship between the desirability of shopping by bicycle and the design of sales outlets. The amount of goods that may be carried on a bicycle is smaller than what could be placed in the trunk of a car. Package size, notable for food and general stores should be adjusted accordingly, and access to bicycle parking spaces for loading the shopping cart contents onto the bicycle should be optimized. This is part of a much more general question of the best way to arrange freight transport in a world on the way to sustainability. Clearly, heavier goods purchased should be delivered to where the customers wants them by the selling agent of shop. All supermarkets should, as some do already, offer to deliver what has been bought to where the buyer lives. For nondaily shopping needs (furniture, appliances, garden utilities, etc.), this is already implemented in many places.

In fact, the rising fraction of shopping transactions made on the Internet is helping to make these adjustments happen. Customers are in the European Union offered far better conditions than in physical shops, including 14 days return right with no questions asked. Delivery to the door is in some cases done by vans belonging to the shop, in other cases by public mail or specialized cargo-handling companies. This type of business is conducted with very large differences in energy efficiency. Some vans drive 80% empty most of the time, even on longer trips, while others are always filled up and serve customers located along a string. This has to do with the amount of deliveries made by the furnisher as well as the skill in circuit planning (which my mathematics friends tell me is not a simple job). There is thus room for considerable improvement, such as by always using electric vehicles whenever cheaper under consideration of all environmental externalities, coordinate public and private delivering services in order to maximize efficiency, vehicle filling and vehicle itinerary. Also, it may be to the advantage of competing services to use common delivery vehicles. In the case of long hauls (for that special sofa produced at the opposite end of the country, and more generally for transport between manufacturer and wholesale agent or between wholesale agents and retail-sale outlets), railway transport should be used in combination with local delivery services. This happens all too rarely at present due to the lack of wish to collaborate and practical problems of train schedules, etc. What have already been optimized are the containers used for bulk transport by ship, train or lorry. These have been standardized and the equipment to easily move them from one vehicle of transportation to the next is available and often integrated into the vehicle. One could go further and suggest that most manufacturers and wholesale warehouses should be located at railway tracks, and that the further reloading places (before getting to retail shops or to customers, employing whole sale or individual packages) have layouts amenable to easy shift from train and large lorries to smaller lorries and vans.

One of the largest consumers of transportation energy is commuting between home and workplace. Earlier nearness of working places or preferred hiring of local people has diminished, partly because more jobs require specialized knowledge and settling patterns have changed. Once, industry was placed in city centres and the population at suburbs (if not also in the centres). Today, many families settle in or near recreational areas (because middle-class incomes allow them to), and many businesses settle away from cities where land for offices and production facilities are cheap. This is the result of the period where even substantial transportation was considered of no importance time wise and of little cost because oil prices were low and environmental concerns forgotten. Is there a solution to this dilemma that has sneaked up behind us?

There are things that may help. Apart from the observation made that an increasing number of jobs may be performed from home and therefore do not require regular commuting to and from the employers workplace, an obvious suggestion is to make it the responsibility of the employer to collect the employee at home and bring her/him back after work. If the time used for commuting in this way is included in the salaried work hours, it is in the employer's interest to make the transport quick and efficient, using different means of transportation depending on the density distribution of employees (buses, minivans, taxi), and in those cases where it is possible, to preferably hire local people. Currently, there is an employee tax credit for commuting expenses in some countries, but this evidently does not provide the carrot for minimizing the cost of commuting in the way that employer administration does.

When it comes to government concerns as to how transportation tasks are performed, there are essentially two ways the society can influence, which options are used and how. One is by imposing a tax progressively hitting the options disfavored, and the other is to introduce regulation excluding the disfavored options. Traditionally, taxation has been used in Europe, while regulation has been used in the USA (for reasons associated with the *minimal-state* paradigm of preventing governments from accumulating money that may be used for social redistribution purposes). The effect of the two avenues is not quite the same. Regulation (whether pertaining to building safe constructions or to energy use) ensures absolute compliance, while as discussed in previous sections, taxation leaves a group of people who do the things they are discouraged from in order to show that they do not care and have plenty of money. Actually, the opposite effect of the one intended may happen, in that the heavily taxed luxury items, such as fast cars, become attractive and status symbols. Therefore, the US preference for regulation would seem the most recommendable route of action, irrespective of how it was initially motivated.

4.1.3 Manufacture and New Materials Provision

The most energy-intensive industries have long been keen to implement measures reducing the energy use, including production line setups with high-temperature heat transfer between processes (cascading) and geometrical arrangements facilitating return of thermal energy to collection points for further use, say, in neighboring industries or district heating. Energy stores can play an important role in reuse of energy within a production plants or making use of transmission networks (Sørensen, 2014).

In both production and raw materials extraction industries, efficiency improvements tend to be associated with invention and introduction of new technologies and new idea for plant design. These improvements are often spurred by criticism for excessive environmental or human health impacts and have proceeded in the direction of eliminating superfluous materials and using less of the necessary materials in the manufacturing processes, making process steps cleaner and using robotic handling of actions that could harm human operators. The production industry is probably more advanced in this respect than the mining industry, whether the latter is working in open fields or drilled underground mines. The future of many raw materials industries is expected to be increasingly influenced by recycling opportunities. The decisive factors then becomes minimizing the work needed for recreating the materials demanded by production industries (e.g., by the reuse options mentioned in Section 3.1.4), exploiting substitution options and making the still required recycling processes more efficient. However, the quest here described for optimization is not currently pursued globally.

Improved recycling efficiency should involve collecting end-of-use products in a form reasonably suited for the separation of components and their reformation into new, pure raw materials (as opposed to indiscriminate incineration, as mentioned in Section 3.1.4). The recommended option of lifetime product service (paid for when buying a product or by lease during the usage period) and manufacturer's product-take-back at end-of-life obligation should enable the collection of statistics helping not only in improving the product by better design and production, but also helping to make the right decisions for reuse or recycling. A key factor will be transportation of end-of-life products for disassembling and formation of new raw materials. Current ideas of globalization (do the job anywhere in the world, where it is cheap) may have to yield, when the true cost including externalities of long-distance intercontinental transportation is considered.

Just-in-time manufacturing means minimizing situations where products hard to sell are gotten rid of below production cost, also in the interest of efficient use of resources. Many products can be sold by a scheme of starting manufacture only when a purchasing order has been received. Generally, the distinction between products with clear mass-production advantages and those with only small volume-of-production cost dependence may change in favor of the latter, due to improved computerized robotic production and the proposed simpler sales methods.

4.2 LIFESTYLE TRAITS

4.2.1 Living in a Community Thriving on Coherence and Learned Skills

Human lifestyles have undergone several changes during the period of increasingly dense living in cities, suburbs and larger size communities with increased interdependence and common regulations affecting individual behavior. Most people still feel a longing for (at least occasionally enjoying) the freedom of not being watched or told what to do by anyone. The security of belonging to an entity of social coherence is another experience, sometimes being felt as attractive and sometimes annoying because it appears to be the exact opposite of living in personal freedom. Clearly, a society where people are close to each other must have rules mutually accepted, in order to allow the high population density without creating frequent conflicts, but first of all, its inhabitants must have a measure of tolerance toward neighbors with possibly different personal preferences.

Many citizens of the current world see themselves a part of a hierarchy of communities. There is family and friends, work colleagues, a local community (sometimes with autonomy in deciding many local matters, say, in city or county administrations), some regional/provincial centres and these combined into a country or nation, which since the introduction of national states have gone from being a tolerated nuisance to an identity-forming entity of the same appeal as the its football team has to many male citizens, at least when they win their games. The transition from accepting to worshipping nations and nationality and flags was introduced as part of the romantic wave sweeping over Europe during the 19th century. With it came new causes to fuel hatred against people hard to distinguish from oneself and to wage wars, as the 19th century national wars and the 20th century world wars show (see Section 4.2.4).

The evidence from Section 3.2.1 suggests that direct democracy, at least before the advent of electronic communication, functioned best in smaller communities. This is also the experience over the last century regarding city and county governments, although the delegates are still elected as representatives. Because people met their local politicians weekly in the local supermarket, there tended to be a toning down of the party-based formal belief sets and an upgrading of the simple goal of improving life and activity within the city or county. A similar closeness to national politicians is rarely felt, and the national government is likely to listen more to industrial and other interest groups than to the somehow distant people voting for a particular government. This becomes worse in cases of poorly constructed supranational entities such as the European Union, and perhaps even more uncertain for the idea of a worldwide governance through an organization such as the United Nations (which, however, still has a certain goodwill locally, due to its fairly objective handling of specific conflict situations such as the events initiating the last war in Iraq).

The suggested conclusion that the previously mentioned remarks may give rise to is that there is a need to distribute governance on different levels, providing decentralized control over issues suitable for that, while maintaining more centralized governance for issues that need uniform solutions, on a national level or on a global level in case of problems such as greenhouse warming.

One may ask whether we need nations at all, noting that the impact they have had since their formation has largely been negative. Could a truncated hierarchy be constructed which retains the local governments for their positive nearness between citizens and their constituencies, but drops the national government that is not anyway felt as representing the citizens, particularly not if political parties are involved, and go directly to the supranational entities that must be there to get coherent solutions to global problems, but of course avoiding distorted solutions where the citizens are deprived influence in favor of special interest groups? The answer is probably that as long as there are nations that would rather extend their territory than see it shrink and where these nations use belligerent rhetoric if not action toward neighbors, it will be difficult to get acceptance for abandoning national entities and the rights they already possess through international law.

Another obstacle to considering more amenable structural divisions of the Earth (e.g., avoiding some presently very big but heterogeneous entities) is the vast differences in political systems prevailing, from dictatorial-controlled nations to countries with some measure of democracy. Of course, the inhabitants of each country have adapted their lifestyles to the prevailing political rules, even if they are not in accord with them, and in many cases, the inhabitants accept the propaganda of the government, accept their country's attacks on neighboring countries and join the rhetoric denigration of those who think differently or choose to live differently, particularly if they inhabit an adjoining nation. While large countries like the United States are the result of meaningful deliberation (plus a necessary civil war), countries like China and Russia for historical reasons comprise populations with very little in common, being annexed during some earlier imperialistic conquests.

Again, to move in the direction of a more reasonable situation, the key factor is knowledge and awareness penetrating to all layers of each society. As regards knowledge, advances are connected with very basic components of lifestyles. Presently, most people get some level of education when they are young and until recently it was believed that the knowledge acquired in this way would be valid during their entire life. Of course, new information is added by mass media and personal contacts, but systematic updating of the knowledge related to work and leisure, to profession and to being a politically active participant in democracy are rarely offered or available. Instead, massive indoctrination is being imposed to condition people to respond positively to advertisements for products or for political parties.

What is needed is a sober educational system offering lifelong continued education for everyone at a level that makes learning agreeable and does not overshoot the target, but still makes sure that the basic requirements for living in a densely populated society are met and that the recipient can both perform a job suited for the person's ambitions and abilities, and participate in the democratic shaping of the society. The tools for such education are simple to establish for people having a job, in that a certain fraction of (paid) time should be set aside for further learning, typically 5–10% (highest for professions such as medical doctors). What is different from current educational updates is that also participation in society and its selected form of democracy should be assisted by the lifelong education, teaching citizens to reflect on questions of governance and not be fooled by propaganda. For people out of work or, in social arrangements that allows it or people who prefer not to work (see Section 4.3.1), continued education must at least contain all the social aspects needed for participation in democratic processes, and knowledge updates that the person chooses to take an interest in. Even if someone is not contemplating to re-enter employment (e.g., pensioners), there may well be skills that they want to acquire, say, in information handling or making the most of recreational time.

A special concern is the education of children. The Greek philosophers discussed in Section 3.2.1 saw a clear need for children not to be educated by their parents. This is no less obvious today, where parents abuse their closeness privilege by imposing religious and political as well as economic views often of a quite fundamentalist nature on their children who are unable and unprepared to question these impacts. In modern societies, the parents' indoctrination is supplemented by kindergartens, preschools/schools as well as by the groups of dominating children present in all such institutions and often representing very backward views on gender issues, on education and on the working of society. The teachers of institutions directed at small children may work to realise more proper proportions in the views of the children attending such institutions, but even here, outdated political views may color the education offered. In other words, achieving the education necessary to participate in a truly democratic society entails a thorough reworking of curricula and teacher education across institutions addressing young children if the archaic views of (some) parents should be seasoned by more accountable information.

4.2.2 Role of Normative Impositions

The UN Declaration of Human Rights gives each individual religious freedom, as long as it does not harm other world citizens, but gives no rights to religious institutions. Yet, most religious practice is currently administered quite harshly by religious organizations, imposing rules penetrating deeply into the daily functioning of the practitioners (and sometimes based on doubtful interpretations of the prophets behind the religion). Furthermore, religious organizations have been fighting each other in a most cruel way, through religious wars, crusades, forced conversion and terror actions, down to harassment of differently aligned neighbors. The Christian church promoted Jesus from prophet to god, as part of their effort to become accepted by the polytheists in Imperial Rome, and further added an elusive ghost called the Holy Spirit, because '3' was supposed to be a more divine number than '2'. As this turned out to be insufficient for swaying the Roman elite, a number of half-gods called 'saints' were added, in analogy to those of the Greek Zeus-based religion. The difference was that while the Greek half-gods were offspring of gods and commoners, the Christian saints were individuals rewarded by the church management, such as Cyril of Alexandria, the murderer of the significant female scientist Hypatia, after she had committed the crimes of being a female scholar and teaching science as an objective subject (Socrates Scholasticus, 440).

Political organizations have similarly in a number of cases interfered in violation of the Human Rights Charter with the daily lives of those adhering to their dogmas, as well as others happening to be under their influence. Examples mentioned in previous chapters are the mob rule of Greek direct democracy in the 5th century BC and of the 1791 French Revolution. Other well-documented cases include the atrocities, including deportation to distant work camps, committed by the Communist Parties of the Soviet Union (not least under Stalin's rule) and China (when ruled by Mao and his wife), for example the murder of hundreds of thousands people during the 'Forward Leap' of forcing farmers to abandon their harvest in order to collect nails dropped on highways for the metal industry. On a smaller, but still significant scale, one could mention the right-wing political parties in a number of countries, who for a long time have prevented medical care from being available also to those not able themselves to pay, and one could mention police actions against labor manifestations and antiwar demonstrators.

It is a responsible government's primary job to make sure that all citizens have optimal conditions for unfolding their talents and desires. This again follows from the Human Rights Declaration. However, not all political dogmas are consistent with this basic rule. For example neo-liberalism is not, because it aims to increase disparity and concentrate wealth. But certain versions of socialism are not either, as they aim at social structures that are not offering maximum personal freedom for as many as possible, in balance with the social obligations also mentioned in the declaration.

4.2.3 Tolerance, Uniformity and Treatment of Minorities

A fundamental condition for making it possible for people to live close together in the type of societies prevailing today is tolerance. Without an attitude of tolerance toward other members of society, any community would break down in a mess of envy and negative communication if not violence and unrest. Tolerance has historically been a characteristic of societies functioning well, but the periods in history where contact was limited allowed societies intolerant toward each other to exist in different locations, with conflicts arising only when exploration, trade or other interaction surpassed a certain level. In today's global society with problems causing mass attempts to migrate to other countries with believed better conditions, tolerance is certainly not the norm. Also in local matters from neighbor disputes to rivalling between professions and trades, the current paradigm of focusing on one's own needs and aspirations has caused the climate for tolerance to decline. Lifestyle variation is no longer seen as a positive trait, but children (and probably adults as well, but more profoundly) are mobbed if they do not adopt the certified lifestyle, where friends are people you have ticked on 'like' at Facebook, and kids not having the latest gadget with a name starting with 'i' are not let into your circles. As noted earlier, selfishness has reached its dominant position along with the neo-liberal economic paradigm, in contrast to the situation earlier in countries passing through a welfare period and having a certain emphasis on at least some features of mutual tolerance.

The way most people use modern technology such as tablets and phones, to play games (often with blind violence) and send totally uninteresting messages and pictures to each other is of course an enormous waste of the opportunity for exploiting the technology to interact intelligently on both a communal and a global level. The software offered or advertised along with any kind of computer often make use of only a minute subset of the possibilities available, and finding more interesting software takes personal initiative and willingness to part from the norms of uniform lifestyles and ignore messages from the antivirus software that what you do is 'unsafe'. There have always been pressures in any society to adhere to the conventional norms of behavior and selection of interests, but this seems to have increased over recent decades, leaving us with less social inhomogeneity than ever before. Lack of conformity is not only a vector for creativeness and inventiveness, but also a vehicle for promoting better understanding among people, both within a society and between societies.

The question of minorities has assumed a very large place in media over the recent decades. Earlier, for example when Hitler came into power in Germany, it was a natural thing for opponents to emigrate to other parts of the world, and particularly to the United States that, being itself a nation built by immigrants, basically received this influx in a positive spirit. This is no longer the case in most countries targeted for migration. People fleeing from brutal dictators and their wars in many parts of the world are no longer welcome elsewhere. They often have to enter other countries illegally, and they are often drowned in unsafe boats offered them at high costs by the mafias already controlling much human and drug trafficking. A counter-argument against accepting these refugees as persecuted is that they may just want to improve their lot in other countries with higher salaries and more opportunities. Certainly, the refugees capable of paying the high fees for being smuggled into another country are not those suffering most under some local upheaval and there would be many left behind who cannot afford the transportation to try to get away. If all those in need should be helped, it would be a mass relocation perhaps beyond the capacity of the richer countries. There are also cultural obstacles to moving people to another physical and social climate, and many refugees do not integrate well in the countries they may manage to reach. It has often been suggested that refugees of local wars and other atrocities should be helped by creating centres in neighboring countries with similar culture and life conditions. However, such centres easily become camps in isolation because the receiving countries are already heavily populated and may have neither spare land nor spare jobs, and in several cases they have problems themselves similar to those of the country being fled from. Furthermore, the international financial contributions often leave much to be desired, even when the centres relieve the donor countries from receiving refugees themselves.

There are no obvious solutions to the refugee problems, other than dealing with the problems people are running away from, and this is currently made difficult on one hand because the United Nations cannot interfere in what is considered internal affairs of a country, and on the other hand, military intervention without UN consent is rarely attempted unless there is a clear economic gain for the intervening country (such as that associated with oil in the Iraq case).

One may go one step further and ask about the situation for minorities within a country. A minority could be a fraction that did not in an election get a majority in parliament but any percentage up to 50% of all votes. It is clearly not in line with democracy based on human rights with its request for tolerance, if a 51% majority governs without taking the conditions of the remaining 49% into account or maybe even prosecute them and rob them of money and assets. However, this is how many countries are governed today, although governments with a bit of decency do not dramatically change the lot of the minority.

The issue may also be taken further 'down' into society, asking questions of tolerance toward any type of minority (immigrants, homosexuals, abortionists, people with long noses or a religion different from the leading one). The democracy of a country must be measured by the way in which all of these smaller or larger minorities are treated. Today we see political parties launching hate campaigns against one or another such minority, without seeing or admitting that this is in contradiction to both the Human Rights Charter and the democracies built upon it. It would seem more than reasonable if a democracy set rules that make it impossible for parties and candidates with stated pejorative views on minorities to gain seats in parliament as well as within the juridical or executive branch of government. The only example of such rules that comes to mind is the German postwar legislation to ban communists and neo-Nazis from any public position.

4.2.4 Resolving Conflicts, Abolishing Wars and Cultivating Peace

Conflicts and wars occupy a central place in the world history of the past several thousand years, and death tolls are frightfully high, not to forget the many mutilated and impoverished people in the areas affected. Figs 4.3–4.5 derive from an attempt to quantify violent death resulting from wars and related atrocities (such as the deaths in World War II concentration camps or mass execution of perceived enemies by dictators). Before discussing the time distribution, a few words must be said about the data sources used in constructing the Figures. Two web pages (Wikipedia, 2015a,b) give quantified death tolls related to wars over the most recent 2000 years, based on written sources providing death estimates. If there is more than one source for a given conflict, a median value is provided. Clearly, the numbers are most trustworthy for recent centuries. Presumable, some wars or war-like conflicts, especially those happening in a distant past and away from the Eurasian area covered by written sources, may have escaped the compilers of these data. Also, some of the ancient history writers may be suspected of having exaggerated enemy death counts in order to please the king or other leader commissioning their studies. This would for instance likely be the case for the Roman war history told by Livius (1 BC) or Plutarch (100).

Going back to the 4th millennium BC, the chronology of wars presented here has been constructed primarily on the basis of Kinder and Hilgemann (1991). This source does not estimate death tolls, but gives a description of each conflict, with maps of land areas involved and the actors intervening in what are often quite complex struggles. Based on these descriptions and roughly known populations in the areas concerned, death tolls have been estimated. In cases where they can be compared with those of the Wikipedia numbers, for example for the 1st millennium after the conventional year zero, the estimated numbers tend to be consistently smaller, perhaps because they are conservative estimates, but perhaps more likely because some of the early Wikipedia numbers, although backed by written documents, are too high, which as suggested could due to a bias at the sources. There are other considerations supporting this observation. For example the number of warriors involved in the Viking raids and settlements all over Europe some thousand years ago are backed by Nordic Sagas giving the number of boats used for the Viking expeditions during the relevant time period (Sørensen, 2012b). The Vikings won some battles and lost others, suggesting that the armies they encountered were of a size not much different from their own, and these observations together thus put an upper limit on the possible death tolls, that is found to be consistent with the estimates made on the basis of event narratives, but not with the largest of the Wikipedia numbers.

An important indicator of war intensity is the ratio of war-induced deaths and the total population in the world, shown in Fig. 4.3. In order to calculate its variation over time, the population data shown in Fig. 4.2 have been used. They are taken from Wikipedia (2015c), based on the US Department of Commerce (2015) and Kremer (1993). It must be admitted,

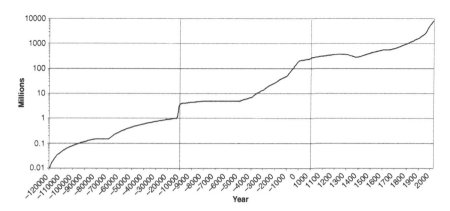

FIGURE 4.2 Estimated world human population development (in millions, based on Sørensen, 2012b; Wikipedia, 2015c and literature cited there). The time scale changes at the vertical lines, from 1000- to 100- to 10-year intervals.

that since both population estimates and war toll estimates are uncertain, their ration is even more so. However, its variations are so pronounced that the interpretation rendered by Fig. 4.3 would appear rather robust. Around 3000 BC, major wars occurred in the Mesopotamian region. The similar battles over Egyptian land appear less deadly, but largely due to the increase in world population taking place during the first 2 millennia BC. Around year zero, however, dramatic increases in war tolls appear,

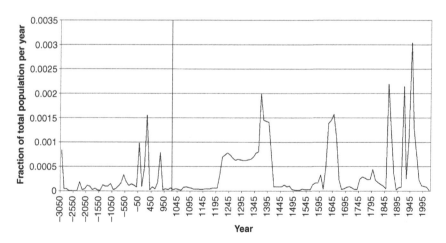

FIGURE 4.3 Estimated global death toll in wars and similar conflicts, relative to estimated total world population. The curve is based on a total of 507 documented war episodes (from Wikipedia, 2015a,b and their references, or based on Kinder and Hilgemann, 1991; using Fig. 4.2 for population data). Unit: ratio between annual deaths averaged over a 10-year period and total world population at the time. The vertical line indicates a change of time scale from 100- to 10-year intervals, this being the spacing between data points plotted.

predominantly due to the imperialistic campaigns of the Roman Empire, but also of the Greek–Macedonian expansion under King Alexander and from 184 to 280 the Three Kingdom War in China. Around 600, the most significant war is in Korea, around 700 the Arab invasion of India and in the mid-8th century an Arab–China conflict. The death tolls are very large, primarily due to the very large populations involved. All the time there are a multitude of wars with smaller death tolls, but according to the data, they amount to only a small fraction of the growing total world population from 940 to 1200, despite incidences such as the crusades of the Roman Catholic Christian church. The subsequent period 1200–1415 is one of the bloodiest in world history, primarily due to the religious Hundred Year War between England and France, the Mongolian conquests and the upcoming of the aggressive Ottoman Empire in Turkey. All assuming that the estimates used are not too far off. The next rise in death tolls occurred during the 16th century, due to religious wars between Christian Catholics and Protestants and also the Spanish invasion of South America and the Japanese invasion of Korea. The following centuries see many devastating wars, including French–English fights in North America, the Napoleonic wars, and conflicts in South Africa, but the most life-consuming wars occur in the 19th century (e.g., the Taiping religion-related revolt in China) and in the 20th century (World Wars I and II, the Sino–Japanese War and the Mao Zedong execution of domestic enemies and his killing of an estimated 30 million people by starvation in 'The great leap forward', mentioned in Section 4.2.2). The Stalin and Mao killings are included although they are not real civil wars (the other side was unable to fight back), but they still lead to as large a death toll as the worst wars and are initiated with the purpose of hitting political opponents within a particular country.

It was suggested earlier in this book, that wars have largely been absent before the invention of property, associated with the transition from migrating hunters and gatherers to sedentary agriculturists. There is a fair amount of evidence for this statement, through analysis of dated skulls and skeletons, for which the start of the agricultural period (around 5,000–12,000 years ago in various parts of Europe and the Middle East) is associated with a marked increase in weapon-induced head and body wounds (Bennike, 1985, for Denmark; see also Becker, 1952; Sørensen, 2012b). Based on such finds related to warfare or at least violent conflicts, one may try to extend the death toll estimates to times before conventional historical source material, as done in Fig. 4.4 (note logarithmic scale). Here, it is also attempted to categorise the wars into territorial wars, religious wars and other kinds, such as civil wars or the one-sided mass killings within countries.

An alternative to the skull-based estimate of violent death during the Neolithic (agricultural) period is to make a time-wise simulation of demographic development based on identified components of birth and death rates. This

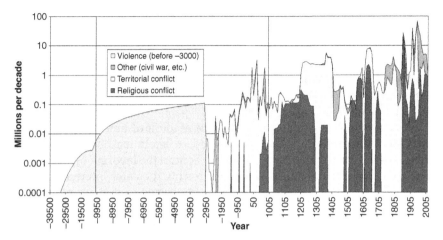

FIGURE 4.4 Estimated global death toll in wars and similar conflicts, divided on main causes. These are stacked with religious causes at bottom, territorial in middle and other causes, such as civil wars, power struggles, etc., at top (from Wikipedia, 2015a,b, or based on Kinder and Hilgemann, 1991). Before the year – 3000, the death tolls by violence in general have been estimated, based on cranial data (Bennike, 1985) and the discussion in Sørensen (2012b). The death toll is given as million victims per decade (note the logarithmic scale), and the time scale changes at the vertical lines, from 1000- to 100- to 10-year intervals.

has only been done for particular cases (Sørensen, 2011b, 2012b), but the finding is consistent with the skull data at the introduction of agriculture. The behavior of the death-by-violence part of Fig. 4.4 is governed by the population growth, with the jump at the introduction of farming as the only large anomaly. Before the year −3000 (i.e. 3000 years before the currently used zero), most violence did not have the character of war, but rather of conflicts on a personal level or raiding a village. After this year, such nonmass violence is not included in Fig. 4.4, although it certainly continued to take place, right up to the gun-shooting incidences in current times. During this long period, war mortality is predominantly among those fighting, although civil population would suffer from the destruction of buildings and food stores, but approaching the present epoch, most wars have very substantial civilian losses.

Fig. 4.4 attempts to divide the wars into territorial wars, religious wars and other types of wars, such as civil wars and power struggles, and is based on the sources mentioned. In several cases, more than one category is involved, for example territorial and religious, and the assignment has involved an estimate of which cause was the leading one. The thesis that religion grew out of mythology was forwarded in Sørensen (2012b), noting that myths were stories elaborated as they were told and retold at village fireplaces, while religion was a mixed bag of dogmas, overviewed by a priesthood, an institution that did not allow practitioners to modify the gospel, even if its origin could have been in mythology. Religions appear soon after the introduction of property and agriculture, when society has started to become stratified and a need for keeping the layers of society at their places is perceived by the ruling classes. Religions often introduce the idea of an afterlife, promised to those toiling with few rewards in their real life. One would then expect religious wars to appear, with some delay, after the Neolithic revolution had introduced agricultural, stratified societies in a given region of the world.

However, it seems that a considerable amount of time had to pass before significant religious wars happened. A likely reason is that not all religions are missionary. As a method to control the layering of a given society, religion does not need war (but possibly it creates internal conflicts in the society). Only when the desire to impose one country's religion on the rest of the world is appearing, does religion become associated with war, notably because the neighboring societies may not want the religion of the missionary society, either because they have their own religion or because they do not want to yield power to foreign, formal institutions. Categorizing the wars through history, there are cases where the label 'religious war' is easy to assign (e.g., the crusades, the Thirty Years' War), while other cases are complex and imposing a new religion on an adversary may only be a 'side effect' of territorial ambitions. Of course, wars of any kind may in many cases be conducted primarily to exhibit the power ambitions of the ruler starting the war, but as often, practical issues such as access to resources play a dominating role in initiating wars.

The first religious war included in Fig. 4.4 is that of the Egyptian polytheists cleansing during the period −1358 to −1350 of the country's followers of the former pharaoh Akhenaten's monotheistic Ra-worship, and then the −845 to −839 Jewish conflicts with adherents of Baal and other gods, and subsequently the −600 to −570 Greek Sikyon–Fokon holy war and the −168 Hasmonean revolt in Israel. During the most recent 2 millennia, clear religious wars emerge primarily from two religions: Christianity and Islam, religions where the requirement of spreading the religion by missionary activity is central. Two periods stand out in Fig. 4.4 as nearly free from religious wars: the 15th century and the period from 1718 to 1850. From 1850, major religious wars are again abundant, notably in China, Africa, India, Pakistan and the Middle East.

Because Fig. 4.4 uses a logarithmic scale in order, the wars with 'other causes' placed on top appear less conspicuous than they are. This is borne out in Fig. 4.5, which reproduce the recent 5000 years from Fig. 4.4 on a linear scale. It shows that civil wars and related atrocities within a nation become significant from about 1740 and are dominating after World War II. Territorial wars appear much reduced after 1950, but some of the internal conflicts actually have the purpose of splitting a given nation into autonomous parts.

In any case, Figs 4.3–4.5 convey the sad message that the human societies are still as belligerent as they were in the late Stone Age and that military

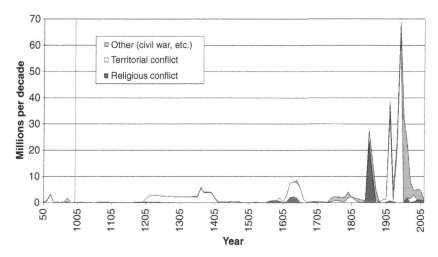

FIGURE 4.5   War tolls and related deaths. Same as Fig. 4.4, but on a nonlogarithmic scale, and confined to the most recent 2000 years. Vertical line indicates change of time scale from 100- to 10-year intervals.

technology has allowed them to increase casualties substantially. Also the nonwar–related violence, depicted in Fig. 4.4 only before the year −3000, does not seem to have diminished. The current level of worldwide average death by nonwar interpersonal violence is at 0.007% of the total population (WHO, 2015), compared with 0.002% of the population at the year −3000, according to Figs 4.2 and 4.4. There are large differences between regions of the world, with currently more than two orders of magnitude differ-ence between the highest and the lowest national figures (World Health Rankings, 2015).

One reason for the recent increase in wars and war-like confrontations is the behavior of the weapons industry, thriving from selling any weap-ons to any state or group willing to pay. The armament of nations, whether peaceful or belligerent, has today reached a level that in most cases is out of proportion to the threats conceivably facing a country. This situation has been ascribed to a military–industrial complex, where the two sectors boost themselves and each other in importance or profits, based on mostly false evaluations of potential threats and certainly false convictions of how to best deal with threats (e.g., when US president Reagan changed the sta-tus of nuclear weapons from tools for deterrence to tools for actual com-bat use, preferably on land far away from his own). The arms trade and the associated lobbyism is presently dominated by Western countries and Russia as sellers, with Middle Eastern, Asian or African totalitarian states and opposition groups including terrorists as large buyers, along with other Western countries. The industrial interests behind this trade have repeatedly blocked any attempt to limit arms trade or at least armament of the most belligerent countries and terrorist organizations, because the current politically accepted neo-liberal paradigm and GNP-fixation does not question any transaction earning money to industry.

A halt to the rise in deadly and disabling conflicts will require action by the world community, actually just enforcing Paragraph 3 in the rules of the UN Declaration of Human Rights (Section 2.3.1). Part of this requirement would be to implement worldwide legislation forbidding any trade of arms, at first between nations but in the next phase also within nations (e.g., outlawing privately owned guns and not making exceptions for self-declared 'hunters' and the like, but only keeping arms permissions for the law-enforcing and defence institutions). It may be argued that this puts the nations capable of producing their own weapons at an advantage relative to countries without this capacity. The truth is rather that the countries not having reached a technological stage allowing them to produce what they see as desirable state-of-the-art weapons are also those most likely to abuse the use of such weapons.

A comprehensive ban on use of weapons and violence to deal with international conflicts has long been a dream, but it requires a supra-national institution such as the United Nations, but upgraded to be capable of stopping any nation trying to break the ban on using arms. This calls for a military role of the United Nations, which today is only indirect, because resolutions by the Security Council to intervene in specific countries must rely on willingness of volunteer nations to put soldiers and weaponry at the disposal of the United Nations. One may ask if it is at all realistic to suggest a future world without wars and aggression. Some would argue that violence is built into the human genes (from our past as hunters and gatherers). I do not think that it can be put as simple as that. Most human behavior is governed by a mix of genetic dispositions and learned attitudes derived from the kind of society we live in and are receiving lessons in 'acceptable conduct' from. Precisely the Mesolithic hunter and gatherer societies were as discussed earlier much less violent than the societies emerging after invention of real estate property, after formation of institutionalized religions and national states, and after the recent predominance of the neo-liberal paradigm, making armed conflicts a desirable way of earning large profits. My conclusion is that things can change, once we realize which ingredients of our societal setup are responsible for the enhanced belligerence. A key component to change would be the nationalistic perceptions surrounding nations (while not abolishing regional differences, just getting away from the romanticizing flag and 'we–them' attitudes that have come to characterize national entities). Equally important is the establishment of an economic framework that does not reward armament and war. The suggestion of replacing the indiscriminate GNP measure of wealth by an MDA (Section 3.1.2) that would recognize some economic transactions as undesirable will be a basic step in this direction.

Those who believe that violence is a genetic trait may consider pinpointing the exact location of the DNA chunk responsible and use genetic engineering (when reaching a more advanced stage than the current one) to modify it. Hopefully, this will long remain outside our reach, as it would equally permit genetic manipulations to make currently peaceful individuals more aggressive and supporters of war as a solution to perceived problems. The alternative is strengthening the international role in governance, by giving an upgraded United Nations the required power to resolve conflicts involving violence between or within nations. Nations must learn to obey an international court of law, and any issue involving more than one nation should by default be referred to such an international body. This does not mean that the international court should interfere in matters clearly of only national or local interest. A basic rule of working democracy is that decisions should be made as closely to those affected as possible. This means retaining local institutions that deal with local issues, national institutions dealing with national issues, but giving the global institutions a more well-defined area of supremacy than at current, and with the power to enforce the decisions it makes. Recent development has gone in the direction of removing power from both the local and the global sphere and giving it to the national governments. Local councils close to the population communities have been replaced by regional entities, where regions are often too large to make people personally know the people making decisions at the regional level. Furthermore, national politicians often unnecessarily mix into local debates that they should have left to the local decision makers (probably as an unbecoming advertisement for national re-election). Also, the media has helped at enlarging the distance between local people and international institutions, by creating a widespread indifference to things happening outside the national boundaries (this is particularly obvious in the European Union, where members of the European Parliament get nearly zero media attention relative to members of the national parliaments). There is a need for creating a novel structure with a clear division of issues between the three levels of governance: local, national and global.

With respect to reducing violence, efforts should clearly be started at a local level, focusing on two issues: to reduce the potential conflict causes (which might be unemployment or economic disparity) and to reinforce tolerance (by quitting the economic dogmas appealing to selfishness, as one obvious issue to deal with). Peace is much easier to create and maintain, if disparities in assets are small, and if local communities retain some measure of decision power.

Of course, there will remain a group of people already educated to hate, to be violent and unforgiving. The goal is to make this group diminish, which almost certainly will require a (maybe prolonged) period with more strict enforcement of the laws most closely related to discouraging violence. This is part of the emphasis on the human rights–human obligations link. Punishments such as imprisonment has in some parts of the world developed into a well-meaning attempt to reintegrate convicts into civil society, but often with less consideration for the victims (in cases of violent crimes) and often creating in-prison gangs planning further crimes and often hardening first-time offenders to a perpetual life as criminals. One (certainly debatable) solution would be shorter terms in prison, but in isolation and with offering of tailored further education rather than entertainment).

4.3 SELECTED SUGGESTIONS FROM RECENT TIMES

This chapter is rounded off by describing some ideas that have surfaced during the recent century. Several of them deal with the financial setup, such as tackling the welfare issue by a basic income, by replacing the growth paradigm by an 'enough-is-enough' paradigm, at least for resource related consumption, and replacing the current bank system that has miserably failed as exposed through recurring collapses and financial crises. Others deal with creating an entrepreneurship capable of handling the many issues still around in the process of reaching sustainability, and finally, questions of controlling the decay of democracy into political party quarrels among career politicians trying to defend some ideology with promises and empty words that are always primarily aimed at being re-elected once more.

4.3.1 Basic Income

Basic income was hinted at in several 16th to 18th century debates, notably by Thomas More (1516) and Ludvig Holberg (1741), and expanded into a theoretically operational form by Bertrand Russell (1918) and by Dennis and Mabel Milner (1918). Russell departs from identifying three major evils of present societies: *physical evils*, such as insufficient food and diseases, *character evils*, such as using violence and *power evils*, based on inequality allowing some members of society to keep the rest down. Poverty is only a symptom; the disease is the slavery imposed by those in power. Russell states that a future society can only avoid the evils by abolishing arms and wars, by redistributing the proceeds of production and service to benefit everyone, but in a way not allowing accumulation of money. Necessities including food should be produced and be freely available, with each individual choosing the mix he or she prefers, education should be enabling not rote learning, and should be free and compulsory for the young, but without any obligation to immediately take a job. A basic salary should be paid regardless of whether or not society can offer work, arranged as drawing rights with a time limitation of a year (to avoid build-up of wealth). There must be communal ownership of land and capital.

The Milners approach basic income from another angle. They are worried about the low productivity of early 20th century British industry, suggesting that workers deliberately slow down their work pace in order not to lose their jobs when the task is completed. Fear of poverty drives the British economy, and one solution would be to introduce a guaranteed basic income, removing the fear and thus opening for improved work efficiency and increasing productivity. The financing of the basic income is suggested to be by a new tax of, say, 20% on all contributions to the national income, entailing the further benefit that everybody would want the total production to increase, no matter whether their share of the cake is only the basic salary available through even distribution of the 20%, or if work activity has made their share larger. All taxes should be collected at the source, and David Milner estimate in his 1920 book that public administration would be greatly reduced.

Evidently, these proposals were influenced by the Russian Revolution and its introduction of centralized communism, but Russell in particular already warned that communism was not the right road, as it would lead to oligarchic concentration of power. Russell wanted as many decisions as practical to be delegated to local communities. In the United Kingdom, labor politician Juliet Rhys-Williams (1943) brought the basic income idea to new political attention during World War II.

The basic income debate surfaced again during the 1970s, particularly in Australia starting with a parliamentary poverty debate (Australian Government, 1975), in the Netherlands (Kuiper, 1976) and in Denmark (Meyer et al., 1978). The official Australian study goes through the quantifications of the need for altered income taxation if a basic income is introduced, much like the one presented below for Denmark, and with several variants. The Danish study by Meyer, Petersen and Sørensen identified a number of problems with the current system, including watering down democracy to party-based politics, unsustainable consumption and increased centralization, as the reasons to seek alternatives, and suggests basic income as a possible solution, along with a discussion of the moral right to work and of worker's influence on production decisions.

During the 1990s and continuing into the present century, while neo-liberalism swept over the broad economic thinking of most countries, its emphasis on increasing inequality has made oppositional ideas of economic justice through income sharing multiply, albeit generally drowning in the noise of the preferred paradigm. In nearly all countries across the world, basic income schemes have been developed and debated, from the United States to Namibia. A modest selection is van Parijs (1992, for the Netherlands); van Trier (1995, Belgium); Rankin (1997, New Zealand); Tomlinson (2001, Australia); Vicenti et al. (2005, Italy); Jordan (2012, the United Kingdom);

Sheahen (2012, the United States); Birnbaum (2012, Sweden); Caputo (2012, several countries); Murray and Pateman (2012, several countries); Lo Vuolo (2013, Latin America) and Fabre et al. (2014, Germany). These studies deal with slightly different variants of basic income, from 'negative tax' to 'guaranteed minimum income', and while the condition for qualifying as a 'basic income' is that no work or other requirements are attached to the payment, there is a considerable number of other studies proposing weaker concepts where the basic income is only for those needing it, implying that the advantage of less administrative work is partially lost, because someone has to decide who is 'needing'. Actually, most countries today have some kind of social payments accorded to recipients found eligible, but unconditional payment to everyone exists only in the common schemes for rewarding fertility by an unconditional fixed payment to the guardian of each child below a certain age. This practice was initiated in countries needing soldiers for their war activities.

Fundamentally, a society consists of its people, their skills and aims, and the stock of buildings, equipment, natural resources, etc. that are available. From one economic point of view, the output (production) of industry and business characterize the level of development, with human work seen as input required along with machinery and capital. From another economic point of view, the basic entity is the population and its aspirations, based on which it may be decided to develop ideas for implementation and to manufacture products seen as desirable, plus eventually other products for export, the earnings of which can satisfy the cost of importing those items best produced elsewhere. The latter approach is suitable for discussing the basic income idea, and it will be further pursued in Chapter 5. Here, a simple model will be presented based on the Danish case study of Section 3.1.3, in order to exhibit the basic changes in a nation's economy implied by the introduction of a basic income. It assumes that the structure of industry and other business is largely unaltered and generate nearly the same revenue as today, in terms of a total sum of money available for salaries, and with roughly the same products and services for domestic consumption and the same import/export relations. A slight reduction in activity is caused by the elimination of some social services and the corresponding financial management, which is not needed when basic income is introduced. In particular, this applies to the setup of social redistribution instruments, modified from that of Fig. 3.5 to one in which the guaranteed basic income for everyone eliminates public unemployment relief and most of the poverty compensation. The objective is to see what burdens the basic income concept may be placing on levels of taxation and chores of administration.

The current situation in Denmark regarding earned and public transfer incomes and taxation for the adult population (4.6 million out of a total of 5.6 million) is shown in Table 4.1 (related to Figs 3.5, 3.7 and 3.8). It is seen that

TABLE 4.1   Personal Income and Taxation of Adult Danish Population 2013

Item	Value
Personal work and private pension income, not public subsidies	29,818
Dividends and interest earnings	1,334
Public transfer incomes	8,206
Personal tax (on income and wealth, not public transfer incomes)	10,582
Tax on transfer incomes (taken as 30%)	2,462
Value added tax (25%) on purchased goods and services	1,218
Point taxes (on cars, alcohol, tobacco, chocolate, etc.)	718

The unit is average €/year per capita (in Aug. 2015, €1 was US $1.096 and 7.45 DKK).
Source: Based on Statistics Denmark (2015a).

the current taxation of the total personal income excluding transfer incomes, directly and indirectly through value added and point taxes, is at about 41%, leaving an average of 18,480 €/year per cap. as disposable income.

In constructing the basic income model, guidance may be derived from the present average (over all adult citizens) private consumption of 9128 €/year per cap. [Statistics Denmark, 2015a; including value added tax (VAT) and point taxes] and from the current subsidy of 3600–5900 €/year per cap. (after tax) to students not living with their parents. For other people losing their employment and having no insurance (most Danes have an unemployment insurance securing 90% of their normal salary for a couple of years), the public cash subsidy is about 10,000 €/year per cap. after tax, but contingent upon active job seeking and diminished if the local authorities do not think the job search is sufficiently wholehearted. In view of this range of figures, a basic salary with no work requirement and amounting to 8000 €/year per cap. is used in the model example, but the effect of alternative levels would be easy to derive. This amount is not subjected to direct tax, but purchases are subject to VAT and point taxes, which are not eliminated because they serve specific behavior-regulating purposes. The basic salaries are paid to all adult Danish citizens, which currently amount to 4.6 million out of a total population of 5.6 million (82%). Table 4.2 shows the new balance for the overall national budget.

The introduction of a basic income is seen to entail some reductions in the need for public subsidy payments, such as government unemployment and sickness salary loss compensation and some of the social subsidies paid to families suddenly having faced a loss of income (e.g., assistance with housing payments), but it does not alter public help not related to job situation, for example child benefit payments. As mentioned, the total national income is not assumed to change

TABLE 4.2   Constructed Basic Income National Budget for Denmark (with Use of Statistics Denmark, 2015a,b)

Item	Current value (2014)	New value
Average adult personal income[a]	33,644	27,000
Average personal income taxation[a]	11,429	11,500
Total taxation (business and private)	23,801[b]	30,300
Basic income payout (not taxed)	0	8,000
Social protection and adjustments (net)	11,459	8,700
Free education, health, security, etc.	13,775	11,600
Public administration	4,250	2,000

The unit is €/year per capita.
[a] Not including basic income, public transfer incomes and rental value of own property.
[b] Including tax on goods and services, but not on social transfers. The government finances in 2014 had a deficit of 5776 €/year per cap., while the budget with basic income is in balance. Without balancing, the basic income case would not substantially alter taxation. The balancing expense is here placed as business taxation.

significantly, considering that there will likely be enough citizens wanting to augment their income by salaried work. A modification in the number of public jobs occurs due to the elimination of several controlling jobs in the public sector, because of the more straightforward payment of the basic salary. Further reductions in total jobs may arise from a streamlining of the financial sector, taking advantage of the simplification of tax payments and public money transfers. Yet, it is quite possible that such 'lost jobs' are compensated by new job-creating activities in other areas, within the public or private sector.

One option for carrying out the optimization of money transactions associated with the basic income scheme will be to replace all paper money with drawing rights on a central bank, which may be seen as a small step from the current electronic payment options, which by the way have been criticized for removing control of money flows from the government, in favor of private banks and little transparent financial institutions. The central bank (which could be the existing national bank, but also a concessioned privately operated bank) would receive all business incomes and pay all business expenses from within the country as well as from abroad, and would transfer net basic and taxed work salaries to the citizen's accounts and taxed dividends to the business owners, and it would handle all trade arrangements (including any import taxation, VAT and applicable point taxes). Taxation would still be progressive but simplified, particularly if a ceiling is placed on maximum income, as suggested. The balance would go to government services without any need for a specific income tax administration, be used for the remaining local or central government services, and for all private or public investments. As this arrangement is likely less labor-intensive than the present setup, a national income reduction has been effected in Table 4.2, associated with the reduction in administrative work. The monetary arrangement described here has further advantages in that tax-fraud, black-market tax-evading work arrangements and other system misuses currently flourishing will diminish or disappear. Finally, the nonsocial government expenditures (education, health, etc.) is modestly reduced, because it includes public works of little sustainable character, such as military adventures abroad (without UN consent) and the continuous adding of motorways between villages and small towns that happens to be in the constituency of a minister or his party chums.

The bottom line balance emerging from Table 4.2 is that the basic income concept may be introduced in a society such as the Danish one with full economic consistency and actually a relatively small change of the current economic flows. The reduction in government expenses is similar to the new basic income expense, so that the level of taxation would be nearly unaltered, except for the assumption to remove the national deficit made in Table 4.2. Around 4.2% of the total personal income in Table 4.2 is dividends and interest earnings from owning businesses or possessing wealth, and the available money for consumption is practically unchanged from the current one (the income is in lines 1 and 4 of Table 4.2, and the after-tax disposable income only slightly above the present one given in line 1 of Table 4.1, while the total private and business taxation in line 3 of Table 4.2 is very modestly increased, when adjusting for the government budget deficit). The basic income idea is thus realisable at the level stipulated, with added benefits in eliminating psychological problems of unemployed people presently having to report regularly to employment centres and possible feeling excluded. Those who take work will actually have a higher income, if the assumption on total national income is maintained, because they get both work salary and basic income. One could express this by stating that the reduction in administrative jobs for the basic income scenario is compensated for by those leaving the work sector and living on the basic income alone. In the further reflections on the implementation of a basic income model one should consider whether work salaries should be decreased due to everybody already getting the basic salary, and it may be contemplated that a ceiling on work salaries could be imposed. A ceiling of six times the basic salary (after tax) will appear generous to most and hurt only a few people currently being paid salaries far above the valuation of any personal effort delivered. Further changes in income and taxation may substantiate if the current GNP-focus is replaced by using an MDA indicator to set goals, which will alter priorities among activities. In any case, such considerations will be left to the scenario discussions in Chapter 5.

4.3.2 Decentralization, Globalization and Continued Education

The claim by Bertrand Russell and others, that decisions should be taken as close as possible to those affected, has given rise to a debate on centralization versus decentralization. Centralization is seen as a way to rationalize governance, a more efficient way to lead a society and thus the least costly form of organizing an entity such as a nation. Against this argument stands the fact that citizens feel less involved in democracy when decisions are taken far away because even if they have had a part in electing the distant rulers, they do not know them or their motivations as well as they know the local members of city or county councils. On the other hand, there are decisions that affect the whole nation or more than that, and which therefore should not be taken (possibly with different outcomes) on a local basis. Curbing emission of gases causing global greenhouse warming cannot be reconciled with one county having a majority for generating energy using coal and the neighboring county by wind.

A simple case illustrating the decentralization issue has been provided by the history of access to electricity. Because many technologies desired by any society are operated on electric power, an effort has been made worldwide to supply electricity to as many people as possible. This was fairly easy in cities and other places with a high population density, but more complex for remote areas. Should one extend the power grids to such areas or find local ways of producing power. Both are expensive. During the early 20th century, alternatives to centralized (usually coal-fired) power stations were diesel sets using oil products, or wind turbines and later solar power. Long power line extensions are expensive, but so is transport of diesel oil and wind turbines, and the choices made in different regions were not always the same, although as time went on, at least until recently, the grid extension was increasingly preferred. One ingredient in this economic argument was the price of centrally produced power.

In early 20th century France, a socialist government decided that all customers should pay the same price for a kilowatt-hour of electricity, no matter if they lived in a city or in the countryside. The city inhabitants thus subsidised the rural users that did not have to pay the full price of the grid extension (Gouvello and Matarasso, 1998). One result was the elimination of a nascent French wind turbine industry. In other parts of Europe, electricity prices varied according to the cost of generation plus transmission and distribution, making electrification of rural areas a very slow process. A third choice, for example in Denmark, was to differentiate prices, but not to the full extent suggested by costs, in order to speed up the electrification of all parts of the country. As a result, the wind industry did well and by 1920 had installed around 90 MW of wind capacity, notably on the many Danish islands, where power grid extensions would have been particularly expensive (Sørensen, 2012b). This eased the acceptance of the current use of wind energy, which by 2014 reached 5000 MW and 39% of the total electricity use (Danish Wind Industry Association, 2015).

The power grid example shows that centralized decisions, however well meant, are not always the best choice. The same can be said of the globalization move that has been advocated in recent decades. In one respect it resembles the French example, as one purpose seems to be to have the same price of commodities all over the world (no customs duties, no taxation on freight). However, the concept is flawed both because 'externalities' are not included and because the quality of the products traded is not considered. The nice words about economic rationality demanding that production is deferred to the regions with the lowest salaries have no validity, when third-world workers are mistreated, when environmentally unsound resources and methods are used in production, when the life of the product is appallingly short and when shipment around the world is done with heavily subsidised prices (e.g., due to the World Trade Organization's unwillingness to consider global climate and environmental impacts of shipping). It is fine that the rich nations want to help the poor ones by deferring production to them, but only if all the other requirements of sustainable production have been met. If not, it would seem better to recognize that the world is divided into regions at very different stages of development, and that the same measuring sticks cannot be employed everywhere.

Another consequence of globalization is that inhabitants of less privileged countries learn about the amenities of the rich (but fail to see that they are unsustainable or see it and want themselves to go through that stage with its large cars) and that many people from the less developed countries want to migrate to the rich countries in order to immediately enjoy their privileges rather than contribute to the slower process of local development. This was not realized then the United Nations formulated it as a human right to travel to and settle in any other country. Presumable the UN lawyers thought of the World War II refugees that were welcomed only in some countries. Clearly, the current mass-migration out of Africa, Mexico and some parts of the Middle and Far East is only partially prompted by the wars in the regions, and as much based on the hope of a better life in Europe, Australia or the United States. Migration around a finite planet will likely always be an inferior solution to making life conditions right where people are.

A decisive factor in most of the issues of globalization and decentralization is the level of education and knowledge from sources other than education. This was discussed in Section 4.2.1, concluding that current societies cannot function properly without continued education of as many of its members as possible. Education that is not rote learning of facts, as in the schools of many countries today, but acquiring skills in problem solving and tools for learning more, including the ability of self-teaching without always needing the presence of a teacher. Such ideas developed from Rousseau (1762) and von Rochow's experiments with teaching young farmers how to do their job better (Alexander, 1918), leading to Grundtvig's Folk High Schools in Denmark (Simon, 1989; Borish, 1991), and more recently received important contributions of problem-oriented teaching and group work for formulation and solution of projects, by Montessori (1912). Her ideas are currently taken up in degrees by a modest number of schools and universities around the world.

The idea behind the Folk High Schools is that education is offered in flexible packages, but without any kind of diploma awarded. It is entirely up to the participant to absorb the knowledge being offered. This does not mean that qualifications should never be evaluated. You would like your medical doctor to know his business, but the point is that exam results obtained at the end of her or his studies (perhaps many years ago) are poor indicators of their present qualifications, regarding familiarity with new scientific advances and the experience in performing the best tests to arrive at a correct diagnosis. Such qualities are best measured by actual performance and can be based on concrete (sometimes statistical) information gathering.

Only with a sufficient level of knowledge can citizens appreciate the complex consequences of globalization based on an uncontrolled trade regime, or the implications of current watering down of democracy by centralized party-rule and efforts to increase the domain sizes for any decentralized local councils, in order to remove the nearness-aspects of democratic participation. A particularly negative example is the European Union that on paper heralds a 'nearness principle' but does not even allow citizens to vote in general referenda on important issues. Also for unfolding the ideas of democratic participation, whether local or for larger regions, educational skills are indispensable, if democracy is not to decay into media and spin doctor–manipulated oligarchies.

4.3.3 Financial and Legal Setup

Many of the observations made in this and preceding chapters regarding the failures of the neo-liberal economic paradigm have been raised in slightly different forms, for example by Merkel (2014) and Streeck (2014). They are also the basis upon which a peaceful protest was launched 2011 in New York (Wikipedia, 2015d), organized by Kalle Lasn and Michael White from the Canadian anticonsumerism magazine *Adbuster*. Out of this grew a worldwide debate on the role of the financial sector in creating the 2007/08 financial crisis and using it to redistribute large funds from common people to the richest 1%. A sober and detailed discussion of these mechanisms has been made by a university based fraction of the Occupy Wall Street movement (Alternative Banking Group, 2013–2014).

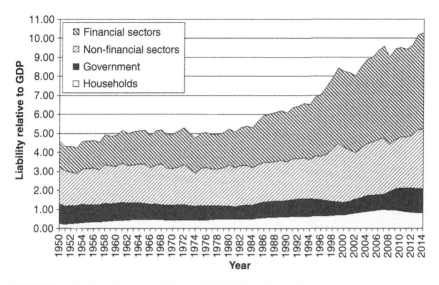

FIGURE 4.6 Development of financial liabilities in the United States, expressed as fraction of the annual Gross Domestic Product and distributed on sectors of the economy. *Based on OECD iLibrary data (2015).*

Highlights include the description of the financial sector developing from institutions helping private and businesses to finance physical commodities like buildings and machinery to half-criminal entities ripping money from those who can least afford it and giving it to the top 1% state of society. Means are emergency loans with annual interest rates often above 100%, but also buying companies in trouble in order to milk them for value and letting them go bankrupt in order to use the apparent losses for the financial company to avoid taxation, and further having betted on the failure in advance (hedge funds). To this comes a wealth of schemes for moving fictive money around, only to create profit for the financial company. Currency exchange including trade aimed at destabilizing weak currencies and making a gain from knowing beforehand that this will happen. Financing political campaigns for corrupt politicians expected to support the activities of the financial sector. Turning mortgage into financial products (derivatives) that will cost the debtor more than anticipated from the original lender. Moving risk from banks to taxpayers, as it happened massively after the 2007/08 financial collapse, but retaining the rewards at the banks. Investment banks that trade in any kind of security, creating suitable crises by themselves and profiting on both positive and negative developments. The growth of the financial sector into such a monster is illustrated in Fig. 4.6 for the United States, where this sector now is responsible for nearly half of all loans and other liabilities, but in contrast to the other sectors depicted, without contributing anything tangible to society.

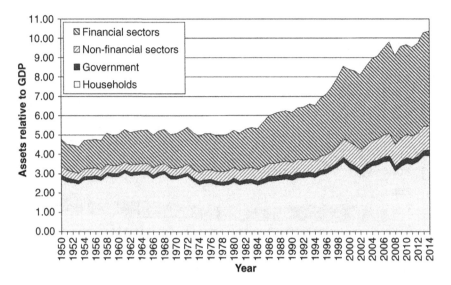

FIGURE 4.7   Development of financial assets in the United States, expressed as fraction of the annual Gross Domestic Product and distributed on sectors of the economy. *Based on OECD iLibrary data (2015).*

The corresponding US financial asset distribution shown in Fig. 4.7 shows how the artificially low current interest rates, caused by the injection of fictive money into the system by the finance sector, has placed more money for consumption in the hands of (the wealthiest) households, while moving the corresponding liabilities to the finance sector (Fig. 4.6) in order to allow the profit to accumulate there. The total balance of the financial sector is still a financial deficit amounting to as much as the US GDP, but figures for assets are less significant than those of liabilities, because the valuation of assets is mostly done arbitrarily by the financial sector (or their stooges in governments and central banks) themselves. The Occupy Wall Street Banking Group has clearly demonstrated how executives move back and forth between the financial predator companies and the government or its central bank.

Streeck (2014) and Merkel (2014) describe this development and argue that it has eroded democracy to such as extent, that there is currently no control of the financial sector, its massive lobbying directed at governments and central banks, and its success in doing all this due to the fact that voters have lost any control over who is making the decisions. The neo-liberal version of capitalism can no longer coexist with democracy, and the nations earlier claiming democratic governance have been turned into oligarchic states that in many ways resemble current China.

Is there a way to revert to democracy and to remove the power of at least the most corrupt part of the financial sector? Streeck hopes that the neo-liberal system will collapse even if there is no clear alternative in sight. The Alternative Banking Group of the Occupy Wall Street movement makes a few concrete suggestions, while realizing that this would just constitute a beginning: Implement a system of responsibility for everyone working in the finance sector, similar to what has worked in the medical profession (the Hippocrates oath, etc.). Outlaw private financial companies' hiring of government regulators, and vice versa. Increase the required cushion against losses that banks are required to have. Require all collateral security to be physical and real. Make regulation affecting ordinary citizens more transparent.

More serious efforts to curb the subversive actions of some players in the finance sector have been touched upon in previous sections of this book. The idea is to revert to the classical role of banks and other financial institutions as mediators of surplus capital to promising business or real estate projects, but to stop any profit-making from empty moving fictive or real money around with no one to gain except the financial company. One may say that this approach is part of an effort to designate certain sectors as being too important to society to leave to vacillating market forces. Such sectors would at least include law-and-order institutions (including police and military), financial institutions, energy providers and communication services, as well as several infrastructure facilitators (roads, transportation systems). Society and its chosen government must have full control of such strategic sector, setting business rules and approving prices. This does not mean (but also does not exclude) government ownership, but the private companies running such services for the society would be what used to be called 'concessioned companies', in that they on one hand may use their business skills to operate efficiently, but on the other hand cannot depart from the rules set by the society in which they operate. The initial rationale for having such concessioned companies was the repeated occurrence of wars that would have disrupted any such service, if it had passed to foreign owners or depended on imported raw materials. One of the rules that should be set by society is therefore that if a strategically important business is owned by a foreign investor, then each transfer of assets of that business out of the country must be approved by the government.

The suggestion was made in Section 4.3.1 that all money-handling operations should be electronic and in the hands of a government bank or a national concessioned financial company. In this way the government would have direct control over the financial sector in a way that would make it straightforward to eliminate any nonpermitted transactions. If also the practice of advertising were abandoned in favor of factual consumer information, the quick loan offers at astronomical interest rates would disappear (in fact, they are already outlawed in some countries). The counter-argument 'what should the poor people do else when they cannot pay their bills' is of course invalid: their situation will in any case be worse if they take a quick loan at an usurious interest rate than if they can settle the issue with the creditor or even if they declare inability to pay. This calls for realizing a third suggestion made earlier in this book, as part of setting new goals and requirements for education: all children growing up should be taught skills in dealing with personal economy and understanding the mechanism of financial transactions.

Considering an augmented role for governments in regulating financial affairs, it is very important that governments can also be held responsible if they do not act in the interest of their constituency. This relates to the democratic setup of the three-partite division of legislative, executive and legal branches. In principle, the legal branch solely judge if the existing laws are complied with, not if they are just or reasonable, and although the executive branch (the government) can suggest news laws to the legislative (parliament) branch, they are only supposed to administer the existing and parliamentary passed legislation. Often, the government tries to assume a more expanded role, where some decisions are going beyond just administering the legislation already in place, and it is therefore important that the members of government can be held accountable. This is in many countries vested with the Supreme Court judges and in some cases it requires a majority consent by parliament to punish or dismiss a minister or president, say, by impeachment. In a few countries, a further avenue has been exploited, by appointing a 'public overseer' (sometimes using the Danish word 'ombudsman', which does not rule out that a woman could hold the job), who by his own choice can raise any issue in parliament that he feels involves a misuse of authority. The overseer is usually a highly respected judge, and the motivation for this additional institution is that the courts cannot raise cases on their own. It is still questionable, if the overseer institution works, except for a certain protection of individuals that may have been mistreated by a government (say by passing laws that are not general but directed at a person or a small group, which is usually illegal). In Denmark, where the 'ombudsman' has been long instated, some governments have ignored his criticism and gotten away with it because they had a parliamentary majority with them. The need for a government to have a majority in parliament is part of the current Danish constitution, but clearly no guarantee that a government may not behave irregularly.

A suggestion may then be that the overseer be given a wider authority but this raises the question of how he/she could be prevented from acting wrongly. Choosing the overseer by direct election does not in itself assure his impartiality and may have to be augmented by more stringent requirements, such as insisting on a spotless record with demonstrated independence from party politics and (at best written) documentation of tolerance and deep knowledge of both laws and the functioning of societies. Rules of fixed duration assignment and no extension should be considered. In the more extreme case where the ombudsman is given the power to oust a government, it would be a very good idea to require the ombudsman himself to subsequently step down and not re-enter politics.

4.3.4 Implementing Change in the Past

As a prelude to suggesting changes in the organization of our societies and their economic rules it may be instructive to recall the tools used in some of the previous cases of major change.

Well-documented examples of changes in human societies can be found in Pharaonic Egypt. Agricultural successes had created luxurious city states with leaders that attained status of gods. This was related to the death culture developed in the region, where even common Egyptians would spend most of their lives to prepare for death and the possible journey to the paradise (located west of Egypt, in the desert), pending upon giving the right answers to entrance questions under guidance from priests and half-gods. How the priesthood invented this intricate system and got acceptance for it remains unclear, but a fact is the strongly hierarchal society with the priests at the top and kings elevated to godly status as Pharaohs. It is therefore remarkable that an attempt to radically change this system was made by the Pharaoh Akhenaten and his wife Nefertiti. They combined a successful foreign policy with the (increasingly forced) introduction of a monotheistic religion that removed power from the existing priesthood. However, in the end the attempt failed and the old regime regained their positions of control. Lesson: changes do not always work out!

More success characterises the developments in classical Greece, where the society was much more open to debate than Egypt and the Middle Eastern countries. Struggles over defining democracy have been described in Section 3.2.1. The introduction of democratic governance was made in steps. One may say steps both forward or backward, considering the cases where changes appeared to have gone too far. In Athens, Solon's first step toward democracy were followed by a direct democracy that developed into a deadly mob rule, but then was modified to keep the positive aspects while avoiding the worst misuse. The rivalling city state of Sparta instead introduced a representative democracy, but in a strange form that also contained elements of Soviet-style forced uniformity. Unfortunately, less written material has survived to describe the Spartan debates that led to this development, compared with the Athenian debates that are generously recorded by writers.

In any case, many of the problems encountered in the Greek wrestling with democracy did resurface in the European Period of Enlightenment, notably during the French revolution that introduced the change through an armed confrontation. These negative events certainly contributed to selecting representative democracy in the United States and also in France once the violence of the revolution had calmed down. These parliamentarian systems soon developed into the party-based version of representative democracy that today has turned into a parody in violation of many of the basic objectives of democracy. The reason this happened is connected with another change happening in Western Europe and its former colonies, namely the introduction of capitalism that basically preferred a dictatorship but increasingly found it manageable to operate under parliamentarian democracy, as long as it could be done through political parties that the financial world was able to control or at least strongly influence. This of course became worse when neo-liberalism appeared on the scene, because as discussed earlier, this variant of capitalism has the demolition of some basic rules of democracy as a basic goal.

The alternative approach of Marxism discussed in Section 3.2 was adapted in Russia and later China, but in a form based on oligarchy or plain dictatorship (Stalin, Mao). Here, the change required a revolution, a conclusion that Marx also had come to, but in Russia introduced with relatively little violence. The Bukunin alternative that developed into European Social Democracies also required a degree of violence, for example between workers' unions and police, but eventually became a vehicle for improving social conditions in most European countries. The high point was probably the introduction of the Scandinavian welfare economies that survived through most of the 20th century despite the Youth Revolt in 1968, dissatisfied with the stagnation of the power distribution in society. Yet the "Americanization" that had started after World War II continued using its advertising tools to penetrate into European consumer preferences and gradually also took over the political scene, where election debates were turned into media shows. As a result, the introduction of neo-liberalism (first in England by Margaret Thatcher and in the United States by Ronald Reagan) went rather unnoticed and unchallenged by a large part of the populations. Party politics had already reduced political debate to media advertising and shows conducted by the party leaders that in some cases were selected for their looks and charming talk or at least had been sent to courses trying to teach them such qualifications.

What would be required for introducing some of the changes suggested in this book? Could it be done through informed debate? Are revolutions still possible? Well, they seemed to be in Libya and Egypt recently, but in the end did not lead to new democracies. Anticipating the criticism that many of the new ideas are not new but were part of the earlier debates as well, a criticism that is well founded, one may first reply that this does not make the suggestions less useful, and secondly that their introduction would not constitute a return to some of the ideologies that we see as having failed. Looking at the manifest of Marx and Engels and the Social Democratic programs described in Sections 3.2 and 2.3.3, one notes that many of the objectives have actually been (or are being) implemented, although not uniformly over all countries calling themselves democratic and particularly not in the countries ruled by religious fundamentalism. This includes:

1. Progressive taxation
2. Government operation of certain strategic sectors
3. Common regulation of the labor market
4. Abolishing child labor
5. Free primary education and most higher education
6. Parliamentary elections with at least option of personal selection
7. A certain decentralization of governance
8. Increasing use of direct democracy by referenda
9. Freedom of the press, in battle with owners
10. Equal rights for men and women
11. Separation of state and church
12. Abolition of death penalty
13. Possibility of free legal representation
14. Free medical assistance
15. A pension system for everyone
16. Unemployment and disability compensation for anyone in need

The first seven of these implemented (although not universally, and for point 2 in some cases again abandoned) points appeared in the Communist Manifesto and points 7–13 in the Tour program of the French Social Democrats, while points 5 and 14–16 are from the Scandinavian Welfare Programs. Not yet implemented, but part of the additions seen as necessary from the argumentation made in this book, are:

17. Abolish the right to turn land into property
18. Abolish inheritance except for memorabilia
19. Government control of banking and loan issuing
20. Forbidding arms trade and warfare
21. Limit terms of political actors and remove party dependence

Of these not-yet implemented changes, the first three were originally suggested by the socialists and the last two by the social democrats. Point 21 requires a reflection on the preservation of cabinet responsibility (that governments should have a majority in parliaments) versus parliament members bound only by their conscience. In the United States, it is accepted that the president and his/her government does not always have a majority in the two parliaments, while in Europe this is not always the case.

In principle, implementing the additions should be possible by peaceful means. It will receive opposition from existing political parties and their backers in industry and financial institutions, but it is for instance difficult to see that governments could not as well have decided to use taxpayer's money to bail the financial sector out with the condition of assuming the control required for making the changes, that are argued in this book to be necessary. If, on the other hand, a revolution turns out to be required, one should worry about the militarization of society involved. A civil war is not the ideal starting point for a new paradigm that includes abolishing wars and weapons. Currently, police have been seen using militant techniques to quench peaceful protests in several countries, governments are using the media to promote their side of the story and to suppress other views, and in some cases (notably in Latin America), the military has on several occasions taken over power. In all these situations, nonviolent resolution of the issues is much to be preferred, but this option has often been deselected, and usually by the established power rather that by the movements for change. Once reforms have been implemented in some countries desiring to establish or return to democracy, it will be easier to spread them to the remaining parts of the world, notably with use of a reinforced international institution of the United Nations type and by applying nonviolent methods such as trade and travel embargos, plus rewards for complying, before as a last resort, military intervention is contemplated.

References

Alexander, T., 1918. The Prussian Elementary Schools. Macmillan Co., New York, Available from: www.archive.org/details/cu31924032699542.

Alternative Banking Group, 2013, revised 2014. Occupy Finance. Available from: altbanking.net

Australian Government, 1975. Poverty in Australia. Commission of Inquiry into Poverty, Canberra (Chapter 6). Available as document HHP011 from: qutvirtual.qut.edu.au

Bahadori, M., 1978. Passive cooling systems in Iranian architecture. Sci. Am. 238, 144–154, Reprinted as: Chapter 9 in Sørensen (2011a).

Becker, C., 1952. Skeletfundet fra Porsmose ved Næstved. Nationalmuseets Arbejdsmark. Årsskrift, Copenhagen (in Danish).

Bennike, P., 1985. Palaeopathology of Danish Skeletons. A Comparative Study of Demography, Disease and Injury. Akademisk Forlag, Copenhagen.

Birnbaum, S., 2012. Basic Income Reconsidered. Palgrave Macmillan, New York.

Borish, S., 1991. The Land of the Living: The Danish Folk High Schools and Denmark's Non-Violent Path to Modernization. Blue Dolphin Publ., Nevada City, CA.

Caputo, R., 2012. Basic Income Guarantee and Politics: International Experience and Perspectives on the Viability of Income Guarantee. Palgrave Macmillan, New York.

Danish Energy Agency, 2015. Building Regulation 2010. Available from: bygningsreglementet. dk/english/0/40

Danish Wind Industry Association, 2015. Knowledge/Statistics/More about the Danish market. Available from: www.windpower.org/en/

Elliott, D., 2013. Fukushima: Impacts and Implications. Palgrave Macmillan, Basingstoke.

Fabre, A., Pallage, S., Zimmermann, C., 2014. Universal basic income versus unemployment insurance. IZA Discussion Paper 8667. Institute for the Study of Labour, Bonn.

Gouvello, C., Matarasso, B., 1998. Bottlenecks and obstacles – elements of a methodology for national diagnosis. In: LTI Research Group (Ed.), Long-Term Integration of Renewable

Energy Sources into the European Energy System. Physica Verlag, Springer, Heidelberg (Chapter 5).

Holberg, L., 1741. Subterranean journey of Niels Klim, containing a new theory of the Earth and the unknown history of the fifth monarchy, extracted from the book collection of the blessed Abelin. Original Latin version published in Leipzig, presumably to circumvent Danish censorship. Danish translations by Baggesen, 1789; and by Zeeberg, 2012, Vandkunsten Publisher, Copenhagen.

Jordan, B., 2012. The low road to basic income? Tax-benefit integration in the UK. J. Social Policy 41, 1–17.

Kinder, H., Hilgemann, V., 1991. Munksgaards Atlas (2 vols.), Copenhagen. Danish enlarged and revised edition based on the German edition Atlas zur Weltgeschichte from 1977 (Deutscher Taschenbuchverlag, Frankfurt).

Kremer, M., 1993. Population growth and technological change: one million B.C. to 1990. Q. J. Econ. 108 (3), 681–716.

Kuiper, J., 1976. Arbeid en inkomen: twee plichten en twee rechten. Social Moondblad Arbeid, pp. 501–512 (in Dutch).

Livius, T., 1 BC, 2001. History of Rome From its Foundation. A translation of the incomplete Latin book Periochae by J. Lendering. Available with comments on www.livius.org/li-ln/livy/livy.htm

Lo Vuolo, R., 2013. Citizen's Income and Welfare Regimes in Latin America. Palgrave Macmillan, New York.

Merkel, W., 2014. Is capitalism compatible with democracy? Zeitschrift für Vergleichende Politik Wissenschaft 8 (2), 109–128.

Meyer, N., Petersen, K., Sørensen, V., 1978. Oprør fra midten. Gyldendal, Copenhagen. English translation, Revolt from the Centre 1981 by C. Hauch at Marion Boyars, London.

Milner, D., 1918. Scheme for a state bonus (pamphlet), followed by Milner, D. (1920). Higher production by a bonus on national output. A proposal for a minimum income for all, varying with national productivity. Allen & Unwin, London. Available as # HHB 213 from: qutvirtual.qut.edu.au

Montessori, M., 1912. The Montessori Method. (A. George, Trans.) (in English). Frederick Stokes Co., New York. Available from: digital.library-upenn.edu/women/Montessori/method/method.html

More, T., 1516. Utopia. Latin original published in Louvain. English translation by Turner, 1963, Penguin Classics, Hammondsworth. Available as E-text 2130 from: www.gutenberg.org.

Murray, M., Pateman, C., 2012. Basic Income Worldwide: Horizons of Reform. Palgrave Macmillan, New York.

OECD iLibrary data, 2015. National accounts/Financial Balance Sheets/720-un-consolidated stocks, annual. Available from: www.OECD-ilibrary.com

Plutarch, M., 100. The parallel lives. English translation from Loeb Classical Library (1920). Available from: penelope.uchicago.edu

Rankin, K., 1997. A new fiscal contract? Constructing a universal basic income and a social wage. Social Policy J. of New Zealand 9, 55-65.

Rhys-Williams, J., 1943. Something to look forward to. Allen Lane, London.

Rousseau, J., 1762. Emile, ou De l'education. (A. Bloom, Trans.) (in English). J. Néaulme, the Hague. Available as E-text 5427 from: www.gutenberg.org

Russell, B., 1918. Proposed Roads to Freedom. Cornwall Press, New York, Henry Holt, London. Available as E-text 690 from: www.gutenberg.org

Sheahen, A., 2012. Basic Income Guarantee; Your Right to Economic Security. Palgrave Macmillan, New York.

Simon, E., 1989. And the sun rises with the farmer. English translation 1998 by K. Parke. Folk School Association of America, Florence, MA. Available from: www.peopleseducation.org/resource/ericasimon.htm

Socrates Scholasticus, 440. The Ecclesiastical History. 7 vols. translated from Greek. Available from: scans.library.utoronto.ca/pdf/2/6/ecclesiasticalhi00socruoft/ ecclesiasticalhi00socruoft.pdf

Sørensen, B., 2007. Assessing current vehicle performance and simulating the performance of hydrogen and hybrid cars. Int. J. Hydrogen Energy 32, 1597–1604.

Sørensen, B., 2010. Renewable Energy, fourth ed. Academic Press/Elsevier, Burlington, MA.

Sørensen, B., 2011a. Renewable Energy Origins and Flows. Renewable Energy Reference Collection, vol. 1. Earthscan, London.

Sørensen, B., 2011b. Demography and the extinction of European Neanderthals. J. Anthropol. Archaeol. 30, 17–29.

Sørensen, B., 2012a. Hydrogen and Fuel Cells, second ed. Academic Press/Elsevier, Burlington, MA.

Sørensen, B., 2012b. A History of Energy – Northern Europe From Stone Age to the Present Day. Earthscan-Routledge, Cambridge.

Sørensen, B., 2014. Energy Intermittency. CRC Press/Taylor & Francis, Boca Raton, FL.

Statistics Denmark, 2015a. Labour, Income and Wealth. Prices and Consumption. Statistical databases INDKP101 and FU5, FU6. Available from: www.statistikbanken.dk

Statistics Denmark, 2015b. National Accounts and Government Finances. Statistical databases OFF12 and OFF29. Available from: www.statistikbanken.dk

Streeck, W., 2014. How will capitalism end? New Left Rev. 87, 35–64.

Tomlinson, J., 2001. Income insecurity: the basic income alternative. Ch. 10 in ebook available at pandora.nla,au/tep/42199

US Department of Commerce, 2015. Historical estimates of world population. Available from: census.gov/population/international/worldpop/table_history.php

van Parijs, P., 1992. Competing justifications of Basic Income. In "Arguing for Basic Income. Ethical foundations for a Radical Reform" (van Parijs, ed.). Verso, London.

van Trier, W., 1995. Everyone a king. PhD thesis, Dept. Sociology, Royal University of Leuwen, Leuwen.

Vicenti, C., Paladini, R., Pollastri, C., 2005. For a wellfare-oriented taxation reform in Italy. Giornale degli Economisti e Anali di Economica 64 (2-3), 189-213.

Weizsäcker, E., Hargroves, K., Smith, M., Desha, N., Rögner, M., Miyake, J., 2009. Factor 5. Earthscan Publ., London.

WHO, 2015. Global Health Observatory Data Repository. World Health Organization. Available from: apps-who.int/gho/data/node.main.CODWORLD?lang=en

Wikipedia, 2015a. List of wars by death toll. Available from: en.wikipedia.org/wiki/List_of wars_by_death_toll

Wikipedia, 2015b. List of wars and anthropogenic disasters by death toll. Available from: en. wikipedia.org/wiki/List_of_wars_and_anthropogenic_disasters_by_death_toll

Wikipedia, 2015c. World population. Available from: en.wikipedia.org/wiki/ World_population

Wikipedia, 2015d. Occupy Wall Street. Available from: en.Wikipedia.org/wiki/Occupy_Wall _Street; see also occupywallstreet.net and occupywallst.org

World Health Rankings, 2015. Violence by country. Available from: www.worldlifeexpectancy. com/cause-of-death/violence/by-country/

5

Implementation of Sustainable Solutions

This concluding chapter will outline a concrete global scenario for a sustainable future, based on several of the suggestions made in the preceding chapters. It is just one of many possible scenarios, but it may facilitate a broad discussion of the issues that need to be dealt with if a sustainable world is to materialize. On each level of accumulation, from local to global, the sections will attempt to describe possible ways of getting started along a sustainable path, with a series of implementation efforts eventually leading to the scenario envisaged (and not stopping there, as further development of society is always required and static societies are definitely not desirable).

The view underlying the scenario presented is that both centralized communism and neo-liberalism have totally failed so that little should be taken over from these paradigms of the past. As bids for a third way, the European social democracies for a while seemed to offer interesting alternatives, and there remains today several positive arrangements from that period, although the parties formed under this banner have moved away from the original beliefs and have merged with the neo-liberal persuasion of capitalism. The scenario outlined here places emphasis on the need for constant reassessment and does not want to be seen as a final solution for the setup of societies. One may describe it as aiming to create the maximum space for free and innovative initiative, while insisting that the results are never permitted to become exploitative. In a nutshell, freedom under responsibility.

5.1 ON THE GLOBAL LEVEL, INCLUDING THOSE ASPECT THAT ARE VALID ON ALL LEVELS

The first serious requirement for getting anywhere is to halt the violence that has characterized much of the previous history of human settlements on the planet. Actually, this was the purpose for which the United Nations

Energy, Resources and Welfare. http://dx.doi.org/10.1016/B978-0-12-803218-3.00005-7

was set up after World War II. The United Nation has not been able fully to live up to the expectation, for at least three reasons. One is that the United Nation has been given no jurisdiction inside nations, where civil wars and genocides by oligarchs and dictations have been all too common. The second is that UN action needs the consent of the Security Council and the General Assembly, which has not always been easy to get because 'friends' of the aggressor country may sit there. The third obstacle is that the United Nation has no military means of its own but has to solicit assistance from member countries to engage in military operations. However, this has generally worked, as for example in the First Iraq War in 1992 against the dictator Saddam Hussein who had invaded Kuwait. More problematic is the first restriction regarding what is called 'internal affairs' of a country. The concern that have made many UN member countries unwilling to extend the mandate of the UN organization to internal affairs is of course the perceived sovereignty of national states that has been interpreted as meaning that any crime committed inside the country is of no concern to the rest of the world. This argument is clearly false, and it appears little credible that a peaceful world can be created without weakening such right to domestic abuse. The large difference in the size of current nations would also seem to be a problem because it appears that the larger a nation is the less willing it is to listen to outside criticism and to international organizations. One way forward would seem to be to weaken the country concept by splitting large countries into independent regions (as it partially happened with the former Soviet Union, although the remaining Russia could also be seen as too large for comfort). The scenario depicted here will deal with governance on three levels: global, regional and local, keeping as much of the constitutional setup common as reasonable, but taking advantage of the options for delegating decisions that do not appreciably affect the next higher level of organization.

The global constitution will be very similar to the existing UN Declaration of Human Rights, but include a few needed updates signalled in Chapter 2. As a constitution, the sections on governance are naturally expanded beyond the hints given in the UN document. Instead of distinguishing between the declaration of human rights and the constitution, they will be merged, as it has already happened to an extent in several countries. This merged version is called a basic constitution, applicable at all levels of societies from global to local, and it may be supplemented by regional and local additions, as discussed in further sections. The universal constitution could contain the following paragraphs (Sørensen, 2012):

PROPOSED BASIC UNIVERSAL CONSTITUTION

1   BACKGROUND
   1.1   The rights and obligations described in this constitution pertain to all human beings, at all ages, independent of gender, outward appearance, lineage and social grouping, and must be enforced without any discrimination.

2 HUMAN RIGHTS

2.1 To cherish, express and live by any personal views, opinions and convictions not harming fellow human beings or diminishing their rights, and not in conflict with human obligations as described later.

2.2 No world citizen can be forced or enticed by any individual or any institution, be it political, religious or of any other nature, to act in specific ways or to adopt or express views that are not part of the basic fabric of human rights, obligations and social arrangements described in this constitution.

2.3 To have access to food, shelter and, when fulfilling human obligations as described later, to human relations and mobility.

2.4 To have fair opportunities for a life with basic welfare, for learning and for enjoying life.

3 HUMAN OBLIGATIONS

3.1 Not to harm any other human being, except in cases of clear self-defence or as part of the legal enforcement of the terms of this constitution. Such enforcement cannot use inhuman or physically harmful methods, except in a sanctioned fight against terror and war caused by agents defying this universal constitution.

3.2 To be considerate of the world's natural makeup, whether the immediate surroundings, the distant or global ones, and whether constituted by living creatures, environment or any other kind of resources.

3.3 To acquire knowledge and skills according to ability and use them to the benefit of society, by actively contributing to the achievement of the common goals.

3.4 To be tolerant and willing to share access to common physical resources and intellectual achievements, limited only by specific exceptions stated in prevailing laws.

4 GOVERNANCE

The constitution is valid globally and thus also for regions and local communities. Additional paragraphs and laws may be added at the regional and local level, in order to allow for cultural differences, differences in financial setup and status, in development and in activity preferences. For this reason, it is not permitted to establish exploiting relations between regions (e.g., for delegating work or for trade). It is allowed to provide assistance across regions (e.g., technical or educational) if mutually agreed.

4.1 Detailed formulations and derivations based upon the general principles stated in this constitution are to be contained in a set of public laws, regulations or singular decisions. These must be of a general nature and cannot target individuals.

4.2   The legislative power to add, edit or delete such laws, regulations and decisions is vested with a parliament elected by general election for a fixed term.

4.3   All members of society can vote in the general elections, after passing a brief test ensuring that they understand the statements of this constitution. Candidates standing for election must have a spotless criminal record and have passed a more comprehensive test of understanding the formulations in the constitution.

4.4   The members of parliament can at most sit there during two terms and are bound only by their own conscience. Premature dismissal of parliament members is possible only after a criminal conviction by the Supreme Court.

4.5   Administration of laws is vested with an executive branch, which is a government elected by parliament and consisting of a prime minister and a number of topical ministers, each disposing over a ministry that employs sufficient staff and possibly extended units (such as police, defence and specialized agencies), in order to be able to deliver an expedient service to all citizens.

4.6   Ministers of the government and their staff (in ministries or extended units) must have a spotless criminal record and have passed the comprehensive test of understanding the constitution. Furthermore, they must have a proven knowledge of the resort area into which they are elected by parliament.

4.7   Ministers and their staff can sit at most during two terms. They can only be dismissed during a term if convicted by the Supreme Court for a criminal offence, or by a vote of non-confidence in the parliament.

4.8   A third of the parliament members, a minister, or 10% of the citizens can decide to have an issue brought to a general referendum among all citizens, the decision of which is binding for the government and parliament.

4.9   Violation of constitutional requirements, laws passed by parliament and specific regulation by governments should be brought before a court of law by a state prosecutor, the office of which is created as part of the executive branch, or by the overseers (see Paragraph 4.13). The court judges will decide on guilt and repercussions, based on fair trials with adequate defence of the accused. Procedures should be stated in appropriate legislation or regulation, honouring principles such as 'innocent until proven guilty', 'equality of all citizens before the law' and 'no conviction can pertain to acts that were not violating any law when they were committed'.

4.10 Legal power is vested in courts headed by judges appointed for two (parliamentary) terms by the parliament. There should be at least two levels of courts in order to make a retrial of a dispute possible. The upper court is called the Supreme Court.

4.11 Judges must, in addition to the extended test of constitutional knowledge, have a spotless criminal record and proven professional knowledge have a record of political independence and service to society.

4.12 Questions about interpretation of the constitution and about laws and regulations can be brought before a court of law by members of parliament, ministers, the overseer (see Paragraph 4.13), or by 10% of the citizens, and by the members of the juridical branch itself. Constitutional questions go directly to the Supreme Court, but decisions require a two-thirds majority to have effect.

4.13 The following overseers are appointed as representing the citizens against suspected misconduct by members of any of the three branches of governance: a secretary of parliament and two ombudspeople, one male and one female, all elected by popular vote (in a referendum) for a period of two parliamentary terms, and having passed the extended constitutional knowledge test. The secretary of parliament is responsible for overseeing that voting and other functions of the parliament are conducted according to rules (including the constitutional rules). The secretary of parliament is staging elections of new overseers. The two ombudspeople deal with worries and with complaints from individual citizens regarding any perceived improper dealing with issues or noncompliance with legal or constitutional rules by the three branches of government. The two ombudspeople should have an office with staff suitable for handling the flow of cases brought before them.

4.14 The ombudspeople can issue appeals to any of the other governance branches and ask them to reconsider any law, regulation or interpretation for fairness and consistency with the constitution, and in serious cases can require one or both the other branches (parliament or judges) to retry the case and ensure that the new outcome is enforced.

4.15 Fraud investigations related to misuse of governance positions can be called for by a third of the members of parliament, or by the ombudspeople. Such impeachment investigations should be conducted by the Supreme Court, omitting the suspected person(s) if they are members of the Supreme Court (and adding judges from other courts if the Supreme Court is in this way seriously reduced). If the case was initiated by the ombudspeople, these should step down and be replaced through a new election, independent of whether or not a conviction is reached in the case raised.

4.16 Changes to the constitution require a parliamentary decision and a subsequent general referendum, both with a majority of two-thirds. The constitution must have been in effect for at least 25 years in order to allow changes.

This constitution opens with a ban on discrimination and continues to list the human rights and obligations, basically copying the UN Declaration but formulated more shortly and sharply. Freedom is restricted only by the condition of not harming other human beings, referring only to physical harm as it is impossible to legislate on the basis of some people feeling offended, which would limit the freedom to nearly nothing. One difference from the UN Declaration is the weaker formulation of the right to free mobility, for example between states and territories, appearing in Paragraph 2.3. The UN Declaration allows not only free movement, but also the right to settle anywhere, in any nation of the world. This is not convenient in an already overpopulated world with large differences in the services offered to citizens in different countries. Convenience refugees, not being prosecuted in their country of origin, but seeking a higher living standard elsewhere, have developed into a major problem. Clearly the more than 3 billion people in the world living below the poverty line cannot all be accepted into high-welfare areas such as the 44,000 km^2 of Denmark and be offered free medical, educational and social services based on taxation of the 5 million people born in this area. Therefore, although mobility is mentioned as generally desirable, limitations may have to be set for permanent resettlement, which should be left to the less general legislation in the regions that for various reasons either want or reject mass immigration. Solving population problems by colonization is a method of the past, as there are no areas left to colonise, except by going extraterrestrial. As mentioned earlier, the new constitution aims at removing problems in the areas that migrants run away from, rather than closing the eyes to local abuse, but it leaves it to the individual regions to move toward a higher living standard, rather than taking the possibly most resourceful citizens away from the less developed regions.

The antiviolence requirement of Paragraph 3.1, outlawing attack wars and armed conflict, naturally has to make the exception needed for enabling the global government to stop regional armed abuse and attacks. However, whether this is best made by a standing supranational army or as now by UN members putting armies at the disposal of the United Nation when requested, is left to be decided among the regions. It may be problematic if only some regions put armies at the disposal at the global level, but there are regions that would prefer not to possess an army (but only police, as currently in Costa Rica), and this is acceptable if such regions instead contribute funds to the global security efforts. Similarly, the merits of regional military or police forces dealing with local conflicts and violent abuses of the constitution, versus only a decentralized local police force, can be left to legislation at the lower levels, but if the regional military or the police oversteps its legitimacy, the global government must be able to intervene, in contrast to the current sanctity of nations. Article 3.1 does only address attacks on humans. More specific legislation should be implemented for avoiding assaults on infrastructure or environment (cf. Article 3.2), ranging from cyber-warfare and physical destruction to threats by computer malware, espionage tools and eavesdropping viruses. For example the 'cloud'-based cyber-attack tools aimed to destroy financial, industrial or energy-handling infrastructure (such as the 'Stuxnet' software allegedly developed by Israel and the United States; Denning, 2012; Broucek and Turner, 2013), which is being prepared by the military of many current nations, should be globally outlawed, except possibly for defensive use by a world government.

The need for protection of the environment as required in Paragraph 3.2 is evident (but absent from the current UN Declaration), and the requirements stated in Paragraphs 3.3 and 3.4 make it clear, that world citizens are obliged to acquire skills and use them for the benefit of society, rather than for the benefit of only the upper 1% elite as with the present neo-liberal paradigm. These paragraphs in essence say that citizens must strive to act in ways governed by the needs of society, rather than just considering personal or employer profit. Specific legislation for this may (or probably 'should' as argued in Section 2.1.1) comprise abolition of land as private property, abolition of descendant's inheritance and restrictions governing copyright and patents. Today, the privatization of land has reached very disturbing proportions, where even streets in cities (notably London) have become privately owned and subjected to special access rules. As regards artists, writers and inventors they should be granted lifelong copyright so that they can sell intellectual property and technical patents and possibly live by it, but multigenerational rights for commercial exploitation unrelated to the creator should not be permitted. Selling a patent or a copyright to someone else should not prolong its period of validity. These views on property differ from those of the UN Declaration of Human Rights.

A community spirit is the foundation upon which the tolerance can be built, that is required for a society whose members depend on each other, and this is the motivation behind formulating the paragraphs on rights and duties in the proposed constitution. The UN Declaration's paragraph on the right to marry freely clearly addresses those societies where this is not the norm, but it is not spelled out here because the general paragraphs on human rights fully covers both marriage and other social rights. Also the United Nation emphasis on the 'right to work' and to form 'labor unions' are considered as little appropriate as a 'right to not work' would be, in the constitution of societies that may choose to institute a basic income strategy, but the UN statements on social security and illegality of discrimination are fully retained through Paragraphs 1.1 and 2.1–2.4.

The governance paragraphs contain some general rules for managing and ensuring human rights and obligations on a global, regional or local level. As in the current setup, the basic tool is a set of laws that regulate what is allowed and disallowed in society, through a set of rules on a much more detailed level than the rights and obligations sections in the constitution, but where global laws are supposed to be sufficiently general to be applicable on all three spatial levels. The preamble of Paragraph 4 forbids more powerful regions to economically or otherwise exploit less powerful ones. Paragraph 4.1 states that laws should be equally valid for all ('equality before the law') and never directed at single persons. The following paragraphs set up the well-known framework of dividing governance on three levels, the legislative, the executive and the legal branch, but adding an independent overseer function aimed to deal with suspected fraud or misuse of power. No kings or presidents are considered necessary because their functions in some current democracies will be covered by the Secretary of Parliament. A few countries already honour the need for overseers, by having appointed 'ombudspersons' that are allowed to bypass the conventional path of going through one of the three conventional government branches, in order to bring an important issue (such as a complaint from citizens, or a governance error) to consideration. The overseer can still decide not to raise complaints if they are found irrelevant. The actual power of the overseers is often unclear. Denmark has implemented such an institution but governments have at times refused to act upon or even listen to a complaint formulated by the ombudsperson.

One should note that the parliament of the proposed World Government is elected democratically, in contrast to the set-up in the present United Nations organization. Other necessary change, if the United Nation shall be transformed into the new world organization, follows from the following paragraphs.

Paragraph 4.3 introduces the new element that citizens should have some idea about what voting is about, at elections or at general referenda. However, no detailed knowledge of the subject matter up for voting is tested, but only the most general notions of the content of the global constitution. Persons not having such knowledge are evidently not able to cast a meaningful vote. The simple test also has the advantage of making it superfluous to put age limits on who can vote. If your 10-year-old daughter passes the test, she is considered fit to vote. Today, bright youngsters having recently made concrete studies of how society is functioning and how it is governed cannot vote, but senile senior citizens can, even if it takes a taxi driver to whisper to them whom to vote for.

In order to run for office, such as sitting in parliament, being a minister or employed by the government, or becoming a judge, Paragraph 4.3 and the corresponding 4.6 and 4.11 require a more comprehensive test of understanding the constitution, plus a spotless criminal record. Today, ministers that have been convicted of fraud reappear as candidates for public office (and not only in Italy), which most citizens find a disgrace. For topical ministers, a further requirement is that they have some knowledge of the field they are about to make decisions for, an obvious requirement that is blatantly absent in the present appointment of ministers only by fitting a puzzle of party coalition distribution of posts (especially in Europe where most governments comprise more than one party) or as a reward for partisan work during election campaigns (USA). Judges should also have some knowledge of law, as already honoured today, but Paragraph 4.11 further requires them to have a record of political independence and service to society, expressed, for example by public debate, written documents or participation in sober voluntary programs. A precise definition of 'political independence' should be established by law.

A further requirement for parliament is stated in Paragraph 4.4 (and 4.7 and 4.10 for the other branches), namely that no one can sit for more than two terms. This aims to prevent politics from becoming a lifetime bread-and-butter profession and reduces the power of political parties that has caused the present disregard and disrespect for the behavior of politicians. A term is in most places currently 4 years, but the duration is here left to be specified by the separate levels of geographical entities. It is not suggested to flatly outlaw parties because people have the right to form associations for discussion, but parties should not control voting in parliament, as this would be in violation of the right of the elected persons to follow only their own conscience.

The administration, or government, consists of a number of topical ministers led by a prime minister, with a staff (ministries, departments and external units such as the police or an environmental protection agency). The ministers are appointed by the parliament, but the prime minister could if desired be replaced by a president, eventually elected by direct voting among all citizens qualified to vote for parliament. No provision has been made to require cabinet responsibility, which is like the current USA, where the president and his administration are often not backed by a majority in any of the two US parliaments. For Europe and other places where governments currently must step down if they no longer have a parliamentary majority, the proposed arrangement will help to reduce the power of parties: The suggested parliament consists of individual persons, and they have agreed on an administration but can vote yes or no to individual proposals no matter who proposes them. It is difficult to see this as a step back because governments are just administrators of existing laws, and it should be the exception that a government sees reason to propose a change existing laws. They may suggest revisions or additions if new topics crop up, and parliament members also have the right to forward new legislation proposals for vote, but it is difficult to see that a government cannot continue to work as long as the parliament does not pass a concrete vote of no confidence, as they of course still can in the proposed scheme.

Paragraph 4.8 introduces the possibility of direct democracy by having general referenda. Rather than making only sitting governments decide on which topics can be put out to direct vote, the possibility of a substantial minority of parliament members or of the population in general calling for a general referendum is opened. This is probably the most important measure made to reduce the current mistrust of party politicians, which as discussed in Sections 2.3.2 and 3.2.4 has reduced representative democracy to oligarchic dictatorships run from behind the scene by small groups of special-interest groups (often large businesses or woolly financial institutions). Neither wanting to have the direct democracy that failed in ancient Athens and less ancient France, nor one based upon a daily push-button vote on any issue, easily leading to populist decisions, this is a suggestion for a balance between direct and representative democracy, a key feature of this proposed constitution but also an obvious point to subject to further debate. The representative members in parliament are elected by voters as persons, but the direct democracy through referenda gives voters a chance to vote on politics.

Additional parliamentary elections within the set period are not envisaged. If an elected candidate leaves the post, the nonelected candidate with highest number of votes may be asked to take over, and so on. Reference to party membership or prearranged coalitions is not allowed in parliament, and a fair media role should be defined through specific legislation, all in line with Paragraph 2.2 (e.g., that journalists cannot be aligned with parties or commercial companies and must display objectivity and report any significant differences in views).

The legal branch is described in Paragraphs 4.9–4.12. Although the equality before the law follows from the human rights, the detailed procedures for accusing individuals, trying them fairly and giving them proper defence assistance must be spelled out in detail. This is here assumed to happen by laws and regulations passed in addition to the constitution because the more detailed such rules are, the more will they reflect the society at a particular point in time, and therefore not be adequate for the general nature of the constitution. This line of thinking is similarly present in current specifications of how the legal branch should go about its business.

A particular dimension of the freedom granted by Paragraph 2.1 is the right to privacy, which has been interpreted as comprising the right to express one-self by phone, correspondence, email and other electronic communication routes, without censorship, listening-in by intelligence agencies or registration by public or private institutions in databases or by other means. The current situation is one where these rights are constantly being violated, and not just in the name of antiterrorist monitoring but also by private companies for allowing targeted advertisement. Selling such tapped information between companies and vendors has become a major business, notably in relation to the Internet, which is about to lose the liberating aspects that were greeted at its introduction. As signalled already in Section 3.1.4, it is difficult to avoid such misuse unless product and service advertising is plainly outlawed (and replaced by an open product and service description with full declaration of positive and negative effects, provided or supervised by the public). If the abolition of advertising is carried through, the worry becomes effectively reduced to illegal registration by government intelligence providers. Recognizing that fighting criminal behavior could benefit from such surveillance (from street web cameras to email monitoring), it is important that the rules for collecting any such information are clear and suitably restrictive, for example that recorded information must only be kept for a finite amount of time and that it may only be used in case a concrete crime is identified. Some nations have already such rules for information registered by public entities.

It is surprising that presently, there seems to be more concern over governments prying into the citizen's private lives (said to be for security reasons) than over commercial surveillance and data collection (e.g., through cookies, collecting Internet IP addresses, and sequences of website visits). This could be interpreted as a subconscious influence from the neoliberal dogma of minimal state and maximum liberty for enterprises in their quest for concentrating wealth. Providers of computer operating systems tend to sell half-finished products, relying on the customers to identify flaws, and excuse their near-daily updates with 'new security threats', while in actuality, this may just tell you that it is very difficult for the system provider to collect personal information from your computer and at the same time prevent other hackers from doing the same.

An important privacy issue is that of voting. Many current constitutions require voting in general elections to be secret, but this may be debated, once a strong focus on the right to express any views not physically harming others is securely implemented. One could argue that citizens voting against spending public money on certain benefits should not receive such benefits if they come into effect due to having been passed by a majority of other voters. But one could equally argue that this is precisely the generosity of a true social community: that voting against a benefit does not entail exclusion.

Some would take exception against the restriction of the right to vote to those having a rudimentary understanding of what the constitution is and say that this is not democracy. However, democracy has never allowed all members of society to vote, and permitting people with no understanding of the matter to vote has precisely been the cause of past and present democratic disasters, all the way back to the historical examples in Greece and France. Present democracies have chosen to weaken the role of the citizens by introducing first representative democracy and then political parties. This is plainly no longer functioning because it has developed into a regime where nations are ruled by lifetime party politicians that repeatedly have ignored opinions held by popular majorities (but which are occasionally revealed in a general referendum). So the present setup is not a democracy except that the term itself is generously used, and the proposed constitution with limited terms for the elected persons and requirements of a basic understanding of what it is all about from the voters should be seen as a bold step in the direction of introducing actual democracy while avoiding mob rule as well as direct and indirect oligarchy.

Paragraphs 4.13–4.15 deal with error and possible misconduct by the members elected for the three branches of governance, and their offices and other staff, in addition to the misinterpretations handled by the courts according to Paragraph 4.12. The handling of such issues is proposed to go through the 'ombudspeople', an extension of a concept developed in the Scandinavian countries. The idea is to have special overseers of high reputation elected by popular referendum (rather than appointed by the sitting government as in the existing Scandinavian cases) and give them the right to probe into behavior and decisions by all three branches of government, and to bring any issue they find questionable to attention and consideration (or renewed consideration) by the appropriate branch of governance. The issue may be a complaint from a citizen of having been the victim of an unconstitutional or unjust decision, an accusation of a civil servant or minister having committed error or fraud, and in the worst case an issue prompting an impeachment charge. There is a fine balance in defining the jurisdiction of the overseers because they should not be given so much power that they can manipulate the government function. For this reason, they are here required to step down after having made a serious accusation (such as an impeachment charge), and give room for election of new overseers independent of the outcome of the investigation called for.

In addition to the two ombudspeople (one male and one female), the secretary of parliament is also an independently elected overseer, charged with ensuring that elections, referenda and parliamentary procedures are complying with the provisions of the constitution. As noted, this setup makes monarchs and presidents superfluous.

Finally, Paragraph 4.16 sets requirements for changing the constitution, demanding a qualified parliamentary majority and a general referendum and that each version of the constitution has had a reasonable time to work before it is changed. Constitutions should not be changed each time a government with a new colour is coming into office, but they may be changed if the fabric of society has changed in substantial amounts, and therefore the minimum period between changing the constitution is set at 25 years. Most intervals between national constitution changes in the past have been at least that long.

5.2 ON THE REGIONAL LEVEL

Regions are national states, provinces in large nations or conglomerates of small states. While the regional governance is generally set up as described in the global constitution (Paragraph 4 and sub-paragraphs), the regional prescriptions would be more detailed, for example by selecting a term length of, say, 4 years for members elected to parliament. The human rights described in Paragraph 2.2 of the general constitution protect individuals from coercion by institutions. It may therefore be accompanied by specific rules or laws setting conditions for establishing institutions and setting limits to their permitted conduct, whether the nature of the institution is political, religious, labor union, professional organization, or another kind. Such legislation should ensure the balance between on one hand the right of anyone to express an opinion and on the other hand the illegality of forcing opinions onto other citizens, not only by physical force but also by threats of reprisal or any missionary or brainwashing-style of influencing. For clarity of competence, such rules should not normally be part of a regional constitution but have the status of specific laws that may be challenged if someone finds them at variance with the global constitution. Similarly, the term length could, as suggested here, be part of the regional constitution, but could also be left to the level of laws, for instance if a region is not sure that the term lengths may not need change.

A specific penal code would be formulated in the form of regional laws, as today, and general remarks on penalty terms and duration may be stated in laws, as guidance to the judges of law. It would be reasonable to expect this code to address violent crimes, theft and fraud, but also to state basic ethical conditions for running or being employed in a business. The current tendency for prisons to become breeding grounds for new criminality could be halted by insisting that all imprisonment should be in isolation. This could be seen as a harder measure and thus be compensated by shorter imprisonments durations, followed by an electronic surveillance period. Punishing crime serves three purposes: (1) making the convicted person realize the severity of the offending act, (2) protecting society from the convict (particularly in case of violent crimes) and (3) preparing the convict for reintegration in society. Retention in isolation accompanied by proper exposure to influence (e.g., through video programs and occasional video conferences with professionals) might be one way of achieving all three objectives.

According to Paragraphs 2.3 and 2.4 of the global constitution, a number of goods must be provided and a number of services offered, either by national companies or by imports, in which case similarly valued products or services for export must be produced. This calls for governance and management efforts to make this economically possible while at the same time ensuring the rights of the citizens. The natural guidance for setting up the necessary system is availability of an indicator such as the measure of desirable activities (MDA) introduced in Chapter 3.

The regional government must therefore set up an office capable of evaluating any activity proposed or ongoing with respect to environmental, resource and health impacts and attach an index of desirability, such as the MDA, to it. How this index may be used to influence the range of goods and services offered will be explained below in connection with the issues of financing and loans. The determination of the MDA, which in some cases involves complex technical measurements and expert judgment, would probably best be conducted by an independent agency reporting to the relevant ministry or government department, where the total statistical data are collected and made public.

The suggestion in the global constitution of demanding a modest test of knowing the content of the constitution as a condition for voting in general elections and referenda may (perhaps only for a transitional period from the current practice of no test) be benevolently implemented as a set of, say, 10 questions posed in the voting box by the electronic voting device, and where the correctness of the replies determine a weight (between zero and one) with which the person's vote is multiplied. This would diminish the embarrassment felt by voters unable to answer all the questions and ensure that every vote counts, albeit not exactly the same (just as today but for different reasons). A sensible rule would be to apply a factor given by the fraction of correct answers squared or raised to the third power. For candidates seeking election or public office (judges etc.) no graduation should be made but the test required to be passed.

The multitude of social services existing today (at least in some countries and with varying conditions) entails considerable workloads for civil servants charged with determining eligibility. As suggested in Chapter 4, administrative simplification could be derived from introducing a flat basic income, the size of which must be determined by the overall financial ability of the regional society (as in the original 1918 Milner suggestion discussed in Section 3.1.2). The basic income could be a sum varying with the MDA.

In order to secure funds for the basic income payments, for government administration and for other public expenses, laws specifying taxation must be passed in parliament. Assuming that the basic income is not taxed, the surplus income derived by citizens accepting work should be

taxed at progressive rates, as they are today in many of the world's nations, say, from 40 to 80%. The progressive tax is supposed to help eliminating some astronomical salaries paid to top executives today, notably by
financial companies and successful industries, and of course to diminish
inequality in society. Examining the extra workloads, responsibilities or
sheer luck (e.g., for a songwriter writing a hit song) connected with the
highest income, it is clear that salaries of say more than 10 times the basic income cannot be factually justified. One may therefore further put an
upper limit on salaries, cautiously (at least at the introduction) set at say
6 times the basic income (also called basic salary) after tax, which for the
income before tax means 30 times the basic income. Few in society, including the most well paid ones, would call this unreasonable.

Currently, transportation to and from the workplace is normally left to
the employee, although working at home is being increasingly introduced,
at least in some countries. Because of the focus on energy efficiency in a
sustainable future, it would be a good idea to further encourage homework
in all cases where it makes sense, and to arrange commuting more efficiently than it presently is. This could as mentioned be by making it the responsibility of employers to fetch and deliver their employees at their home,
of course with an adjustment in salary reflecting the employee's saving
in transportation outlays. In practice, employers or transport companies
working for several enterprises would use efficient routes of transportation,
from large buses and trains to minibuses and taxis or bicycles, depending
on the location of the employees. A company would be encouraged to use
neighborhood workforces when feasible, and as the employer is paying for
the transport, it is not likely that solutions enlarging the commuting time
will be used, so the employees would not have to fear excessive transport
times due to say a large number of stops. Of course, employees would have
to be ready to be picked up at the agreed time.

All members of society are supposed to derive their income by a salary, basic or rewarding work. Businesses can of course come out with any
positive or negative result, but the people associated with the businesses
are in any case paid for their work by a salary. Today, there is also the possibility of earning money as dividends on an invested sum (shares traded
on stock markets, bonds etc.), but the analysis of the financial sector made
in Chapter 4 suggests that this is not the best way of financing the activities in society that are most needed and most useful to the citizens.

The alternative brought forward in Section 4.3.3 involves removing
cash from companies and citizens, making all financial transactions electronic and going through a central banking institution (run by the government or by a concessioned enterprise). All incoming and outgoing payments, whether for individuals or for companies, pass the central bank
and is subjected to (at least annual) audit by an independent auditor (who
should be more independent than the auditors and accountants required

of businesses today, which are usually employed or hired by the company itself). Having one account for each person or enterprise means that over-spending is only possible with an established credit arrangement and that loan procedures should be initiated before the balance becomes negative. For a business with a large positive earning, the amount stays on the company's account, but using it for new investments would be checked by the central bank, largely ruling out indirect payments to individuals or spending on inferior tasks (bribing, fraud, purchase of luxury cars and restaurant visits are among those currently flourishing in certain companies).

As for the cases where an enterprise needs a loan in order to carry on, or to achieve a desired expansion, the central bank will evaluate the application through an independent committee, charged with supporting the investments that contribute most to the MDA of the region. Adherence to the success criterion of maximizing MDA imply that empty transactions of financial manipulation and product advertisement (especially for products that are not attractive by themselves*) must be banned and that all manufacture and service adhere to strict rules of high efficiency, low material's use and minimal environmental and health impacts. Companies could be guided in this by formulating the most commonly applicable requirements in laws and regulations.

Legislation and regulation is a fair way of achieving sustainable solutions because they apply equally to all enterprises and thus do not disturb or alter the competitive balance (which is why regulation such as the building code example in Section 4.1.1 and several US examples has worked much better than the taxation of 'incorrect behavior' introduced in some European countries on the basis of the misconception that the market will solve all problems, if prices are adjusted a bit to reflect some of the 'externalities'). The legislation and regulation regarding sustainability of products and services naturally apply not only when a company seeks a loans or wants to use a surplus for investment, but applies equally to all products and services offered for sale in the region. Ensuring such compliance up front helps avoid producing inferior products that subsequently have to be retracted from the market, but of course some negative issues will escape the checking and first appear after the product has been sold. For this reason there will still be a need for consumer protection by warranties, and as mentioned at best for the entire life of the product, provided conditions for maintenance are fulfilled by the user. For companies repeatedly sending products not complying with the environmental, resource and health requirements on the market or not honouring warranties, legislation should allow revoking the permit of such companies to sell products (that is closing the business).

---

*As my grandma used to say, if a product needs advertising, there must be something wrong with it!

Advertisement should, as suggested, be replaced by a consumer information service set up by the government and run independently from manufacturers and sellers. It would be Internet-based and contain all factual information of the product, including energy use and environmental impacts, provided by independent testing outfits (such as TÜV, a group of independent consultants engaged in testing products, i.a. for health and environmental impacts, followed by conditional issuing of a certification. It has branches all across Germany; see Wikipedia, 2015), and it would be financed by a fee from the companies wanting to bring a product to the regional market. The fee could reasonably be set at 5% of the retail price of the item tested, although some tests could be more complex and require a higher fee. Current advertisement costs are from about 5% to over 50% (the highest ones for undesirable and health-damaging products such as most cosmetics), so the consumer information service solution will generally lower the prices of goods. Some similar product information services already exist today, but are usually operated by private businesses and include only products from sellers paying for the service.

While price setting can generally be left to the market, there would be strategic sectors where price setting should be approved by the government (through the relevant ministry/department). This is to ensure that the basic income method is not sabotaged by price fluctuations on a volatile market. If the price fluctuations are coming from imported goods over which the government has no control, a subsidy may be needed in order to keep the consumer price acceptable for basic income-only earners.

Personnel working in the police and (if deemed necessary) defence forces should in addition to the extended constitution test have a high level of education with emphasis on the psychology and humanitarian aspects of carrying out acts of violence on behalf of the state, in a society where violence is otherwise outlawed. Rules of conduct for the two exceptional institutions allowed to exert violence should be established by law. The global constitution calls for abolition of arms and war as a response to political problems. In particular, the nuclear arms that serve no purpose other than mutual destruction, with the current arsenal deployed in a risky manner wide open to errors (Anthony, 2013), must be removed and destroyed, and at the opposite end of the conflict scale, ordinary citizens must be deprived of the right to purchase and carry weapons. This would best be achieved by outlawing any weapon's trade, within or between regions, and by placing all current nuclear weapons under control of the global government, until they can be safely destroyed.

Innovation is essential for any society. Regional research and development efforts should therefore be established and the current ones encouraged, in private industry and business as well as in public support organizations such as universities and technology advisory organizations. This differs from the present situation only in that research in a sustainable society cannot be proprietary. Good ideas should of course be rewarded, by offering competitive advantage and patent protection, but only for a limited time because otherwise technological progress would be hampered by the associated barriers facing other inventive individuals and enterprises. Proprietary rights including patents should expire after say one year, which would put the originating company in a fair position to be first with a marketable product but not quench subsequent competition to the benefit of the consumers. Today, patents are sometimes taken out with the sole purpose of preventing others from using an idea, and with no intention of immediate use. Openness and publication of achievements and experiences should become the norm.

In order to encourage a business environment of progressivism and renewal, it is important that all people from floor to management receive continued education. This could be through courses or time accorded to self-study, and a reasonable duration would be at least 3 weeks a year. For service sectors such as the medical sector, continued education is even more essential, and no hospital or practicing doctor should be able to avoid spending a reasonable time (say, a month a year) on further education. The health sector must be set up to allow such updating of knowledge, whether it is achieved by setting aside time to follow new professional literature, by courses or by other means. For reasons of patient security, the continued education of health sector personnel should be monitored by suitable tests or exams. Similar levels of continued education should be compulsory in several other sectors, with verifying tests in case of professions with direct impact on health and safety of citizens and more relaxed verification in other areas.

Educational offers should be made to satisfy the requirements of society. This means that interdisciplinarity and cultivation of ability for self-study must be essential throughout all basic education, and that further study should rely on these qualifications. Admission conditions for the higher levels of education should make sure that a sufficient number of students choose the subjects that society cannot be without, even if it means limiting access to popular studies for which few jobs can be offered. Proper advice, which is often absent today, will help achieve this.

An incomplete list of some of the items discussed is presented, in the form of a regional constitution, but with some provisions that could be moved to the less fundamental level of laws. The paragraphs are numbered from 5 and up, in order to avoid confusion with the global constitution.

A REGIONAL CONSTITUTION

5   The economic setup in the region should encourage free enterprise with environmental and social responsibility and should ensure a fair distribution of the advantages and proceeds from such activities.

   5.1   Perennial natural property such as land cannot be privately owned but must be leased from the state through a designated government authority for a duration not extending beyond the lifetime of the renter, who has the full obligation to maintain the property without deterioration to the end of the lease.

5.2   Descendants have the right to inheritance of minor personal items of affection. The parliament will through legislation set rules and limitations that regulate more substantial amounts of inheritance, and that apply taxation to such inheritance.

5.3   A Central Bank shall be set up to handle all economic transactions including basic income, salaries, construction, machinery procurement, trade, import, export, public or private services and social subsidies.

5.4   Each citizen and each business enterprise shall have an account in the Central Bank to handle income and expenditures and to accumulate surpluses. The bank shall establish an independent professional evaluation council for examining loan applications and issue recommendations.

5.5   The government shall set up an independent office to estimate the environmental and social desirability of all economic activities (published as an MDA), and the parliament can set conditions for enterprises based on the MDA contribution of their production of goods or services, concretizing the general statement at the opening of Paragraph 5.

5.6   A common nontaxed basic income shall be transferred to every citizen without any obligation attached. The amount shall be set by parliament and shall, as a rule, be adjusted according to the total MDA of the region.

5.7   Salaries paid for work of any nature are taxable income and the parliament shall set a progressive scale of tax to be deducted at the Central Bank from salaries paid by enterprises or other employers before transfer to the personal account of the salaried person.

5.8   Income (personal or business surpluses) may be used for investment. The nature of substantial investments shall be evaluated (by the office described in Paragraph 5.5) and approved, as required for loans according to Paragraph 5.4.

5.9   A property valuation office should be set up with aims similar to the activity valuation office described in Paragraph 5.5. Properties of finite lifetime should be evaluated with consideration of the service rendered, working life duration, lifetime energy use, health and environmental impact. The ratings made by the office should be made publicly available.

5.10   New equipment produced or imported (buildings, vehicles, machinery etc.) should have the highest rating in the manufacturing and product valuations described in Paragraphs 5.5 and 5.9. Detailed criteria should be made available in building codes, vehicle standard requirements etc., that are regularly updated.

5.11 For property with long life (such as buildings), the government may use the rating described in Paragraph 5.9 to set requirements for upgrading at the event of owner change (an example is the energy rating of a building, which may be required by upgrade or rebuilding to reach given levels).

5.12 Prices of consumer goods are set by the manufacturer and seller, except that the government may set maximum prices for strategic goods (such as certain food, clothes and energy goods) that are affordable to citizens with no income above the basic income, and may decide to contribute a subsidy if there are just causes that industry or importers temporarily cannot meet the price requirement for the essential products.

5.13 The government should establish a web-based information service giving detailed and reliable information on quality and usage, prices and the impacts of all consumer goods and services licensed for sale, evaluated as described in Paragraph 5.9 and allowing the consumers to compare similar offers. No direct or indirect advertising of products or services is permitted, and the sellers are required to pay a fee for this government information service. The privacy of individuals using the Internet services must be respected and personal information collected only with permission by concrete legislation.

5.14 The government shall operate or control a public media enterprise with presence on popular media such as television, radio and the Internet, broadcasting news from both the region and the rest of the world, material for understanding the activities of the parliament and the government and the issues brought up for political debate, documentaries on a broad range of subjects, cultural transmissions covering the entire range of activities from art and music to films, dance and theatre.

5.15 Establishment of further privately operated media is welcomed. They do not like the public have the obligation to transmit materials needed for the democratic processes to function, but they can do so if they like. They must obey the ban on advertising and use subscription to finance their productions.

5.16 All media, whether written, televised or offered by or for electronic devices (including games) must show objectivity in any discussion of both political and other nonfiction issues. Violence cannot be shown in any of the media without a factual context.

5.17 Journalists working for any of the media must in addition to their professional education pass the same test and character requirements as those imposed on judges of law in Paragraph 4.11 of the global constitution.

5.18 In line with the global human right to adequate shelter, the government should maintain a backstop building enterprise capable of delivering affordable housing to everyone, including citizens with only the basic income. If this requires a subsidy, private building enterprises should get the same subsidy on actual projects, in order to preserve fair competition.

6 Election of candidates for parliament and other functions to be carried out by elected persons should follow the requirements of the global constitution, with the precisions stated below.

6.1 Members are elected to parliament in their own capacity. Political party alignment is not permitted and as stated in Paragraph 4.4 of the global constitution, the members are during their term responsible only to their own conscience. Any political party is considered a private organization, allowed by Paragraph 2.1 of the global constitution if it heeds Paragraph 2.2 and does not exert any undue influence on parliamentary members or affairs.

6.2 The term length for parliament is 4 years, and citizens elected for parliament can sit at most for two terms. The overseers (the ombudspeople and the parliamentary secretary) are elected for two terms, that is 8 years, and cannot be re-elected.

6.3 A standing election committee is appointed by the secretary of parliament. It establishes procedures for parliament elections and for general referenda, such as for electing overseers. It also formulates the questions in the minor and major constitutional tests. Criticism of the functioning of the election committee should be directed to the ombudspeople.

6.4 The government is responsible for staging the elections and referenda required by the global constitution upon request from the global government, for the citizens of the region, and to faithfully report the results back.

7 The social services of the government, including as a minimum health, education and social safety net functions, are to be provided with adequate administrational offices.

7.1 A government branch administering the health sector is created, with responsibility for setting up a high-quality public hospital service with coverage throughout the region, and a practicing doctor service with uniform rules but administered locally.

7.2 A government branch administering the educational sector is created, with responsibility for setting up all levels of education, from primary schools to secondary and tertiary educations of university or vocational nature, as well as provisions for continued education for people working in business, service or other sectors. Opportunities and obligations for receiving continued education should be specified by legislation. Primary education and part of continued education may be decentralized to local authorities. Admission criteria for specific tertiary studies may be set by government.

7.3 Uniform minimum curriculum requirements for each kind of educational offers should be established by use of professional committees. These requirements should be formulated in broad terms, leaving reasonable choice to each educational institution.

7.4 Government should establish research institutions for basic research as well as industrial innovation and topical research, generally honouring the tradition for combining research and education in universities and other such institutions. Specific legislation should be set up to limit the duration of patent- and copyrights in such a way that social progress is not hindered by private interests.

7.5 A government branch responsible for initiating, administering and supervising public works (including the building activity described in Paragraph 5.18) should be established, dealing with infrastructure, communication, energy provision and transmission/distribution, transportation (road or track-based land transport, sea and air transport), buildings etc.. Some projects may be deferred to local administrations.

7.6 A social safety net administration branch in government is established, responsible for helping all people needing special care (such as disabled people and people otherwise incapable of taking full care of their lives). Collaboration with local centres is recommended.

8 Handling of any violation of this constitution according to Paragraph 4.9 in the global constitution and its specifications in special laws is to be detailed in a penal code, with indictment and court judging arranged by offices of the legislative branch of governance, and assistance from the police force as described below in apprehending suspects.

8.1 The government should form a police force charged with responsibility for maintaining general order in society, for avoiding violence and for apprehending violators and bringing them before a court of law within 24 h. All police personnel must have passed the extended constitution test,

have no criminal record and have received tertiary education
with emphasis on psychological and humanitarian aspects
associated with carrying out acts of violence on behalf of
the state, in a society where violence is otherwise outlawed.
The police institution further operates prisons for offenders
convicted at court, according to rules set by law.

8.2   Chemical and biological weapons are not allowed and firearms
only in the hands of police and constitutionally formed military
forces. Any illegal arms existing among citizens must be
handed in to the police for destruction.

8.3   The government may form a military defence if it sees external
threats that cannot be dealt with by peaceful means. Military
force may only be used against aggressors from outside the
region, or by request from the global government following
acceptance in the global parliament.

8.4   It follows from Paragraph 3.1 of the global constitution that
weapons of mass destruction may not be possessed by any
region, and regional governments are thus charged with
destroying those existing in a planned process approved by the
global government.

The regional constitution calls for several laws to specify details. However, it should be considered a virtue to have as few laws as possible. Current national states often have a bewildering number of laws, and the governments formed by political parties seem to take pride in having as many laws and law revisions passed as possible during their terms. In reality, few developments inside or outside a region call for laws beyond those already needed to implement the provisions of the constitution, and it should be a rare event that a government or a parliamentarian proposes a new law to parliament.

Adherents to the liberal persuasion may dislike the role that the regional constitution gives to government in approving products and setting rules for consumer protection, although this is only an extension of regulation existing in most countries. They may also point out that parting with useless and detrimental products and services will lower the total activity level (gross national product, GNP) and make many unemployed. In reality, producing only needed and useful products, and meaningful services, will permit the required working hours to be reduced, which most citizens will welcome, and although it is true that the GNP will diminish, the MDA will likely rise and more people will be able to satisfy basic needs and engage in desirable activities. Still, the possibility of unwanted unemployment exists in the period of transition from the present to the new system, and this is precisely why the basic income is essential in order not to lose anyone during the change of conditions.

The regional constitution contains in Paragraph 5.13 confirmation of the right to privacy in relation to any kind of advertising or data collection by private entities. However, the possibility of governments collecting information on citizens is left but has to be based on specific legislation and court order. This is already a hotly debated topic. Presently, there is open data collection of personal data in several areas (health records, tax matters, dwelling properties etc.) and covert collection of data globally by the intelligence offices of nations such as the United States. The problem is that such data could be useful in preventing certain crimes, but also present several risks of misuse. Many current national states have created detailed legislation for protecting personal data, and the problem is that they are not always followed, and particularly not internationally, due to the complexity of cross-border facilities such as the Internet. This could be altered by the introduction of an international government with extended authority.

A particular issue of privacy is the secrecy of voting, introduced in many present nations with representative party democracy in order to prevent punishment of people voting against the winning party, once this party is in office. Enforcing Paragraph 2.1 of the proposed global constitution would in principle remove the need for protecting voters in this way, and, as examples, fighting violence and terror could benefit from knowing who voted against gun control laws, and providers of strategic electricity supply could benefit from knowing which people welcome solar panels and wind turbines in their neighborhood and which did not want energy installations 'in their backyards'. In any case, the stance taken here is to leave such discussions to regional and local attention leading to regional legislation, where in the energy example, limitations on sustainable energy installations has to go together with extra stringent legislation on high-energy efficiency.

5.3 ON THE LOCAL LEVEL

The near society should be a community with high levels of interpersonal acquaintance, such as villages, small cities or small communes with mixed settlements. In recent decades, many of these have lost their local governance provision because several local areas have been merged into one, changing the relation between citizens and the new 'local' government by creating a distance that party-politicians see as convenient for introducing party politics at the local level. The earlier local units were characterized by politicians and administrators concerned with local wellbeing, often forgetting the party quarrels characterizing national politics.

It may not always be needed to formulate constitution extensions at the local level, but if made, they should be seen as concretisations of the Global and Regional Constitutions. Local governance is about making sure that local inhabitants can satisfy basic needs such as food and a dwelling that can be afforded from the basic income alone. The role of the local government is to establish the facilities for this or see that the private sector does, but also to assist citizens in selecting social arrangements permitting maximum welfare. One component of this may be communal living in families or larger groups, taking advantage of sharing dwellings, food preparation and various kinds of tools.

The local government will also, as today, be in charge of primary education, that may be compulsory for a longer period than presently, say, until age 18, but with options for more levels and skills aimed at, from book to practice-oriented study. Continued education should also be encouraged and monitored by the local government, in collaboration with regional institutions charged with the set-up of different types of continued education. The duration of educational extension attendance could as suggested in the regional constitution be set at least 3 weeks a year, with option to lump together, say if an individual wants to move to a new profession. Assistance for bringing up babies and small children would as now be through establishment of nurseries and kindergartens offering enabling pedagogy, play and increasing elements of learning. Indications of parental influence in directions incompatible with the constitution should be followed up with dialogue, and in case this has no effect by reporting to authorities.

Possible paragraphs in a local constitution may (with numbering continued from the regional constitution) be as follows.

A LOCAL CONSTITUTION

9   Public services suitable for local administration should be managed locally but in close collaboration with the corresponding regional administrations. Selection of local government and administration follows the general rules (at global and regional levels) but require local ties of the elected and preferably also of the hired persons.

   9.1   A network of practicing general or specialist doctors and medical centres should be made available locally, providing health service nearness to users. Citizens may freely choose which services to use, and the cost of medical consultation and care is paid by public funds. The local government sets rules and fees for those providing medical services, including medicine, subject to approval by the regional government with transfer of associated funds to the local administrations.

   9.2   Nurseries, kindergartens and primary schools should be established by local governments as close as feasible to population agglomerations, using funds transferred from the regional government. Safe commuting to these facilities for children should be ensured through offering transportation by use of communal vehicles, detached bicycle paths etc. Continued education institutions may also be established locally, if their functioning is thereby improved.

9.3   Some of the social safety net institutions outlined in Paragraph 7.5 of the regional constitution may be operated by the local governments. This should be agreed between the local and regional levels of government according to the advantages that may be obtained by decentralization.

9.4   Public works not affecting other local communities should be administered locally (area planning, buildings, including checking of building code compliance, and establishment of cultural activity centres, minor roads, bicycle paths and walkways or trails, for example).

9.5   As a rule, pollution-producing activities including polluting manufacture should not be allowed in cities and villages. During a planned transition to 100% pollution-free vehicles, combustion engine vehicles may not enter cities and larger villages (say, above 4000 inhabitants) and the local government should establish sufficient parking space outside the city for visitors, with options to rent bicycles, e-bikes, pollution-free trolleys and/or public transport for people unable to bicycle and cargo e-trolleys for people wishing to transport large loads. Public transportation by road or track should use pollution-free vehicles. Pollution-free delivery vehicles may be allowed to enter the city during certain hours of the day.

9.6   The local government is responsible for making any local industry or other business conform to the rules on health and environmental impacts set by the regional government according to Paragraph 5.5 of the regional constitution. For example all food production could be required to be done according to the ecological (organic) prescriptions for production, and limitations on additives should be applied to all products coming into contact with or being ingested or absorbed by human beings (food, cosmetics, clothes), for example by conforming to a 'positive list' (formulated on the regional or global level) of permitted substances, as is presently the case for food alone.

9.7   Local governments should have an amount of funds transferred from the regional level to stage cultural activities in the local area.

9.8   The local government is responsible for making recreational facilities available to citizens (from forest to beach, if applicable), with proper access routes (roads, trails) in accordance with Paragraph 5.1 of the regional constitution. Access to all types of land (which according to the regional constitution cannot be owned) is free but tourists should show consideration when passing by agricultural fields or inhabited private buildings.

5.4 IT CAN BE DONE!

In the past, changing economic paradigms has usually required an armed revolution, even in cases where the new paradigm has embraced a quest for abolishing war. Can a scenario such as the one described above be introduced peacefully? I think so. It ameliorates the living conditions of 99% of the population, and although the remaining 1% may oppose it, using propaganda through political parties and media they own, I believe that the current educational system, however imperfect, together with common sense should tell the other 99% to vote for change, if only an opening of media plurality and emphasis on real democracy could be created. To accomplish that should be within reach of a dedicated minority, acting through peaceful manifestations like for instance those of the Occupy Wall Street movement. Details such as how to regain strategic enterprises previously privatized can be handled, for example by passing legislation narrowly defining the obligations of providers of services such as energy, telecommunication or financing. If the privatized companies accept this, all is well and they function like consessioned companies. If they do not, they are breaking the law and face criminal charges and confiscation of their properties.

Yet the opposition cannot be underestimated. Financial crises like the 2007/8 one that should have brought about change, but have instead been swept under the rug by the neo-liberal power junta. This is possible as long as ordinary people have no power. If they protest their job is taken from them and they are reduced to nothing in the present system. The initial protesters may come from an academic elite with little power and their preferred weapon of public discourse requires access to media. The Internet is losing, or has lost, its liberating aspects and is controlled by advertising agents, making sure that serious debates are drowned between the 'like' and 'don't like' buttons accompanying content on many sites. My container of hope appears to be perforated in the bottom! Must we really have major catastrophes occur before a majority will wake up, and then maybe too late?

References

Anthony, P., 2013. The man who saved the world. Documentary on the imminent 1983 global nuclear war prevented by colonel Stanislav Petrov. Description available from: themanwhosavedtheworldmovie.com

Broucek, V., Turner, P., 2013. Technical, legal and ethical dilemmas: distinguishing risks arising from malware and cyber-attack tools in the 'cloud'—a forensic computing perspective. J. Comp. Virol. 9, 27–33.

Denning, D., 2012. Stuxnet: what has changed? Future Internet 4, 672–687.

Sørensen, B., 2012. A History of Energy—Northern Europe from Stone Age to the Present day. Earthscan-Routledge, Cambridge, A constitution like the proposed one was first published 2005 as a discussion paper at http://energy.ruc.dk; under 'Downloads'.

Wikipedia, 2015. Technischer Überwachungsverein. en.wikipedia.org/wiki (in English).

Subject Index

A

Abolishing wars, 149
Adbuster magazine, 166
Administrators, 107, 109, 185, 200
Advertisement. *See also* Campaigns
Advertising agents, 14
Advertising campaigns, 56
Agriculture, 5, 6
  climate for, 10
  crop yields in north European
    societies, 12
  history, 6
  practices, 18, 19
  reforms, 12
Air pollution, 1, 14
Alternative Banking Group of the Occupy
    Wall Street movement, 168
Aluminium mines
  operating, survey of the global
    distribution of, 22
  resources, 23
Animal food, 28
Animal husbandry, 11
Annual average solar radiation, on
    horizontal plane, 25
Anti-violence requirement, 182
Antivirus software, 148
Architecture, 133
Aristotle, 109
Arms reduction, 119
Athenian debates, 171
Athenian democracy, 109
Auditors, 191
Average income per capita, in Denmark, 83

B

Bacteria, 29
Bakunin, Mikhail, 106
Basic income, 158–161, 163
Basic income debate, 159
Basic income model, 161
Basic universal constitution, 178
  background, 178
  governance, 179–182

human obligations, 179
human rights, 179
Bicycles, 140
Biodiversity, 14, 28
Biofuels, 27
Biomass, 27
  geographical distribution of net
    production, 28
  production, 28
  resources, 28
Bourgeoisie, 104
Bream, 12
Buildings
  cooking ranges and appliances for, 137
  cool and frost compartments, 138
  Danish building code
    alteration, 135
    prescriptions for a new 200-m^2 floor
      area, 136
    weakness, 134
  Danish regulation
    incorporating energy efficiency in, 134
  energy labelling system, rated and
    labelled A–F, 137
  energy use of, 132
  heat loss through windows, 136
  heat transfer, 134
  insulate the shell, 133
  materials, 37
    used for, 134
  one-family detached house, 135
  proper design, 134
Business environment, 194
Buying and selling goods, 96
  advertising, 98
  certified life-cycle analysis, 98
  current legislation, 97
  database information, 98
  end-of-life handling of products, 98
  environmental issues, 97
  goods categorised as, 96
  guarantee periods, 99
  improved resource and energy
    usage, 99

Printed in the United States
By Bookmasters